SpringerBriefs in Optimization

Series Editors

Panos M. Pardalos
János D. Pintér
Stephen M. Robinson
Tamás Terlaky
My T. Thai

T0215707

For further volumes:
http://www.springer.com/series/8918

SpringerBriefs in Optimization showcases algorithmic and theoretical techniques, case studies, and applications within the broad-based field of optimization. Manuscripts related to the ever-growing applications of optimization in applied mathematics, engineering, medicine, economics, and other applied sciences are encouraged.

Leonidas S. Pitsoulis

Topics in Matroid Theory

 Springer

Leonidas S. Pitsoulis
Department of Mathematical,
    Physical and Computational Sciences,
    School of Engineering
Aristotle University of Thessaloniki
Thessaloniki
Greece

ISSN 2190-8354   ISSN 2191-575X (electronic)
ISBN 978-1-4614-8956-6   ISBN 978-1-4614-8957-3 (eBook)
DOI 10.1007/978-1-4614-8957-3
Springer New York Heidelberg Dordrecht London

Library of Congress Control Number: 2013947069

Mathematics Subject Classification (2010): 05B35, 52B40, 68R10, 97K30, 90C27

Printed on acid-free paper

Springer is part of Springer Science+Business Media (www.springer.com)

To Lenio, Vasilina, and Alexios

Preface

The purpose of this monograph is to expose a less-known decomposition result in matroid theory that provides a structural characterization of graphic matroids, and show how this can be extended to signed-graphic matroids. The immediate algorithmic consequences of the decomposition are also examined. In order to make the exposition self-contained we also provide a brief, but nevertheless solid introduction to the elements of matroid theory, by presenting the way it exhibits itself in three different contexts, namely graph theory, vector spaces, and transversal theory. This book is intended for graduate students and researchers from graph theory, operations research, and combinatorial optimization, who are interested in theoretical and algorithmic applications of matroid theory.

Acknowledgments: This work has been partially funded by the European Union (European Social Fund—ESF) and Greek national funds through the Operational Program "Education and Lifelong Learning" of the National Strategic Reference Framework (NSRF)—Research Funding Program: Thalis. Investing in knowledge society through the European Social Fund.

Thessaloniki, Greece, March 2013   Leonidas S. Pitsoulis

Contents

Figures

Symbols

$(E, \mathscr{I})$	Independence system
$\alpha(G)$	Edge connectivity number of a graph G
$\cap_i M_i$	Intersection of matroids
$\kappa(G)$	Vertex connectivity number of a graph G
$\kappa(M)$	Vertical connectivity number of a matroid M
$\lambda(G)$	Connectivity number of a graph G
$\lambda(M)$	Connectivity number of a matroid M
$\langle X \rangle$	Span of a set of vectors X
$\mathbb{N}$	Natural numbers
$\mathbb{R}$	Real numbers
$\mathbb{Z}$	Integer numbers
$\mathbb{Z}_+$	Non-negative integer numbers
$\mathscr{B}$	Family of bases
$\mathscr{C}$	Family of circuits
$\mathscr{I}$	Family of independent sets
$\pi(M, B, Y)$	Partition of Y as determined by B
$\Sigma(G, \sigma)$	Signed graph Σ with underlying graph G and sign function σ
$\Sigma / \{e\}$	Contraction of $e \in E(\Sigma)$ in signed graph Σ
$\Sigma \backslash \{e\}$	Deletion of $e \in E(\Sigma)$ in signed graph Σ
$\dim(V)$	Dimension of vector space V
$\mathbf{e}_k$	Unit vector
A/Y	Contraction of $Y \in columns(A)$ in matrix A
$A \Delta B$	Symmetric difference of sets
$A(:, j)$	j-th Column of matrix A
$A(i,:)$	i-th Row of matrix A
$A - B$	Difference of sets
A^T	Transpose of matrix A
$A_{\overrightarrow{\Sigma}}$	Incidence matrix of a bidirected graph $\overrightarrow{\Sigma}$
$A_{\overrightarrow{G}}$	Incidence matrix of a directed graph $\overrightarrow{G}$
A_{Σ}	Incidence matrix of a signed graph Σ
A_G	Incidence matrix of a graph G
$C(B, v)$	Components determined by bridges
$cl(X)$	Closure of a set X

$columns(A)$	Column indices of matrix A
$d_G(v)$	Degree of a vertex $v \in V(G)$
$G \cap H$	Intersection of graphs G and H
G/Y	Contraction of $Y \subseteq E(G)$ in graph G
$G.Y$	Contraction to $Y \subseteq E(G)$ in graph G
$G \cup H$	Union of graphs G and H
$G \backslash X$	Deletion of $X \subseteq E(G)$ in graph G
$G|X$	Deletion to $X \subseteq E(G)$ in graph G
$GF(2)$	Binary field
$GF(3)$	Ternary field
I_n	Identity matrix of size n
k_G	Number of connected components of a graph G
$K_{n_1, n_2, \dots, n_k}$	Complete k-partite graph
K_n	Complete graph on n vertices
$lr(X)$	Low rank of a set X
M/X	Contraction of elements X from matroid M
$M.X$	Contraction to elements X in matroid M
$M \backslash X$	Deletion of $X \in columns(A)$ in matrix A
$M \backslash X$	Deletion of elements X from matroid M
$M|X$	Deletion to elements X in matroid M
$M(E, \mathscr{F})$	Transversal matroid of set system $(E, \mathscr{F})$
$M(E, \mathscr{I})$	Matroid M on E with independence family $\mathscr{I}$
$M(G)$	Graphic matroid of G
$M[A]$	Vector matroid of matrix A
M^*	Dual matroid of M
$M_1 \cong M_2$	Matroid M_1 isomorphic with M_2
$N(A)$	Nullspace of matrix A
$N(A^T)$	Row space of matrix A
$N_G(v)$	Neighborhood of v in G
o	Orientation function of a graph G
$R(A)$	Column space of matrix A
$r(G)$	Rank of a graph G
$r(X)$	Rank of a set X in independence systems
$r(X)$	Rank of a set of vectors X
$R_{G_1} \circledast_Y R_{G_2}$	Star composition of G_1 and G_2 in Y
$rows(A)$	Row indices of matrix A
$S^\perp$	Orthogonal complement of subspace S
$U_{k,n}$	Uniform matroid on n elements and rank k
$Y(B, v)$	Partition of a cocircuit

Chapter 1
Introduction

Matroids were initially conceived as a generalization of graphs and linear independence in vector spaces, and they were introduced in the seminal paper by Whitney (1935) where he laid the foundations of the core in matroid theory. Whitney (1935) demonstrated the existence of equivalent axiomatic definitions, which is a characteristic feature of matroids, and established fundamental properties such as representability, duality, and connectivity. The first set of basic structural results on matroids appear in a series of papers by Tutte (1956, 1958a, b, 1959). Tutte provided characterizations for several important classes of matroids, such as graphic matroids, matroids representable over the binary field, and over any field, and in doing so he expanded the theory by introducing notions such as higher connectivity and the theory of bridges. The connection of matroids to optimization was established by Jack Edmonds by recognizing that matroids can be defined algorithmically by the greedy algorithm, and showing a number of important results on matroid partitioning and intersection, polymatroids, and submodular functions (see Edmmonds (1970, 1971); Edmonds and Fulkerson (1965)).

1.1 Past Literature

Although a complete bibliography for matroid theory is beyond the scope of this monograph, we will provide a representative list of books and chapters that have appeared.

There are a number of books written on matroid theory. The earliest book on the subject seems to be by Crapo and Rota (1970), where matroids or combinatorial geometries as the authors called them at that time, were presented in a lattice theoretic framework. Another early book on matroids is the book by Tutte (1971), which is a collection of lectures that the author gave at RAND Corporation in 1965. The book by von Randow (1975) can be described as a general introductory monograph, with emphasis on the equivalent axiomatic definitions of matroids. One of the first

L. S. Pitsoulis, *Topics in Matroid Theory*,
SpringerBriefs in Optimization, DOI: 10.1007/978-1-4614-8957-3_1,
© Leonidas S. Pitsoulis 2014

books aimed toward undergraduates is the book by Bryan and Perfect (1980). The book by Welsh (1976) can be considered as the first comprehensive textbook on matroid theory, since it contains almost all aspects of the theory up to that date in a single volume, along with a number of exercises. The most comprehensive book on matroid theory for researchers and graduate students is without doubt Oxley's treatise (Oxley 1992), and its second expanded edition (Oxley 2011). The book by Truemper (1992) deals primarily with the notion of decomposition in matroid theory, and it can also serve well as an introductory text. The books by Recski (1989) and Murota (1999) are concerned mainly with applications of the theory to systems analysis and structures. Finally, the textbook by Gordon and McNulty (2012) has a strong geometric emphasis and is directed toward undergraduate students. There are also three edited volumes devoted to matroid theory by White (1986, 1987, 1992), which contain contributed chapters by leading experts in the field on most aspects of matroid theory.

A number of expository papers and book chapters have appeared on matroid theory. One of the earliest survey papers is by Tutte (1965), which some readers may find hard to follow due to the author's unique writing style and notation. Wilson's exceptional survey paper (Wilson 1973) presents an approach for illuminating the unification properties of matroids, that we also adopted in Chap. 2. The chapters by Welsh (1995), Seymour (1995), and Bixby and Cunningham (1995) appear on the same volume edited by Graham et al. (1995). Most of the chapters on matroid theory appear in textbooks on Combinatorial Optimization or Combinatorics, such as Lawler (1979, Chaps. 7, 8, 9), Korte and Vygen (2001, Chaps. 13, 14), Lee (2004, Chaps. 1, 3), Aigner (1979, Chaps. 6, 7), Schrijver (2003, Chaps. 39–49), Nemhauser and Wolsy (1989, Chap. III.3), and Papadimitriou and Steiglitz (1982, Chap. 12).

1.2 Preliminaries

The set of natural numbers $\{1, 2, 3, \ldots\}$ is denoted by $\mathbb{N}$, the set of integers by $\mathbb{Z}$, the set of non-negative integers by $\mathbb{Z}_+$, and the set of reals by $\mathbb{R}$. While familiarity with basic set theory is assumed, here we will just state some frequently used notations. All the sets considered in this book are finite, unless otherwise stated. The number of elements in a set A will be denoted by $|A|$. The **power-set** of a set A is the set of all subsets of A, including the empty set and A, and we write 2^A. We denote with $A - B$ the deletion of B from A, that is, the set which contains the elements of A which are not in B. The set $\{A_1, A_2, \ldots, A_m\}$ of nonempty subsets of A will be called a **partition** of A if the A_i are pairwise disjoint, and their union is A. Given two sets A and B, we will say that A **meets** B if neither one is contained in the other and $A \cap B \neq \emptyset$. The **symmetric difference** of two sets A and B is defined as

$$A \bigtriangleup B = (A \cup B) - (A \cap B).$$

For a set A and some index set I, by $(a_i : i \in I)$ or $\{a_i\}_{i \in I}$ we denote the **family** of elements in A indexed by I, as defined by some mapping $\phi : I \rightarrow A$ where $\phi(i) = a_i$. Note that a family is not set, since order and multiplicity of elements are of relevance. For the family of subsets of some set A we will write $\mathscr{F} = (S_i : i \in I)$, where the corresponding mapping will be $\phi : I \rightarrow 2^A$. We will refer to the tuple $(A, \mathscr{F})$ as a **set system**. Given a set system $(A, \mathscr{F})$ a subset $X \subseteq A$ is **maximal** with respect to $\mathscr{F}$, if $X \in \mathscr{F}$ and there does not exist $Y \in \mathscr{F}$ such that $X \subset Y$. Moreover, $X \subseteq A$ is **minimal** with respect to $\mathscr{F}$, if $X \in \mathscr{F}$ and there does not exist $Y \in \mathscr{F}$ such that $Y \subset X$.

A matrix A with n rows and m columns with elements a_{ij} over a field $\mathbb{F}$ will be written as $A = (a_{ij}) \in \mathbb{F}^{n \times m}$. The i-th row of A will be denoted by $A(i, :)$ and the j-th column by $A(:, j)$. By $rows(A)$ and $columns(A)$ we mean the index sets of the rows and columns of A respectively. We will use column notation for vectors, which will be written using boldface letters. So by $\mathbf{x} \in \mathbb{F}^n$ we mean the ordered n-tuple of elements from $\mathbb{F}$, arranged in a column. The identity matrix will be denoted by I_n, and is an $n \times n$ matrix with ones in the diagonal and zero anywhere else. When the dimension of the identity matrix follows from the context the subscript n will be omitted. The unit vector $\mathbf{e}_k$ is a vector with zeroes everywhere except at position k that has a one. We write A^T for the transpose of a matrix A.

We will deal with matrices with elements over the field of real numbers $\mathbb{R}$, the binary field $GF(2)$ and the ternary field $GF(3)$. The binary field has only two elements 0 and 1, and the operations of addition and multiplication are performed modulo 2 as follows:

$$
\begin{array}{c|cc}
+ & 0 & 1 \\
\hline
0 & 0 & 1 \\
1 & 1 & 0
\end{array}
\qquad
\begin{array}{c|cc}
\times & 0 & 1 \\
\hline
0 & 0 & 0 \\
1 & 0 & 1
\end{array}
$$

The ternary field has three elements 0, 1, and 2, and the operations of addition and multiplication are performed modulo 3 as follows:

$$
\begin{array}{c|ccc}
+ & 0 & 1 & 2 \\
\hline
0 & 0 & 1 & 2 \\
1 & 1 & 2 & 0 \\
2 & 2 & 0 & 1
\end{array}
\qquad
\begin{array}{c|ccc}
\times & 0 & 1 & 2 \\
\hline
0 & 0 & 0 & 0 \\
1 & 0 & 1 & 2 \\
2 & 0 & 2 & 1
\end{array}
$$

The following operations on the rows of a matrix over a field $\mathbb{F}$ are called **elementary row operations**:

(i) Interchange two rows.
(ii) Multiply a row by some nonzero member of $\mathbb{F}$.
(iii) Replace a row by its sum with a multiple of another row.

If matrix A is obtained from B by elementary row operations, then we say that A and B are **row equivalent**. By **pivoting** on an element a_{ij} of a matrix A, we mean the application of elementary row operations (ii) and (iii) in order to make $a_{ij} = 1$ and all other elements of column j zero. Given a matrix $A \in \mathbb{F}^{m \times n}$ we can define the following three fundamental subspaces:

$$\text{null space} \quad N(A) = \{\mathbf{x} \in \mathbb{F}^n : A\mathbf{x} = \mathbf{0}\},$$
$$\text{column space} \quad R(A) = \{\mathbf{y} \in \mathbb{F}^m : A\mathbf{x} = \mathbf{y}, \text{ for some } \mathbf{x} \in \mathbb{F}^n\},$$
$$\text{row space} \quad R(A^T) = \{\mathbf{y} \in \mathbb{F}^n : A^T\mathbf{x} = \mathbf{y}, \text{ for some } \mathbf{x} \in \mathbb{F}^m\}.$$

If S is a subspace of a vector space $\mathbb{F}^n$ the **orthogonal complement** of S is defined as

$$S^\perp = \{\mathbf{x} \in \mathbb{F}^n : \mathbf{x}^T\mathbf{y} = 0 \text{ for every } \mathbf{y} \in S\}.$$

The following fundamental theorem from Linear Algebra relates the column and row spaces of a matrix.

Theorem 1.1 *For a matrix A we have $N(A) = R(A^T)^\perp$ and $N(A^T) = R(A)^\perp$.*

1.3 Organization of the Book

We will use the symbol $\square$ to indicate the end of a proof or example. Text in **boldface** indicates a new term definition which almost always is associated with an entry in the index at the back of the book, while text in *italics* indicates a notion that needs to be emphasized. We make extended use of figures to illustrate concepts, especially since most of the concepts presented in this book are graph theoretic.

The book is organized into six chapters as follows. Chapter 1 is this introductory chapter. Chapter 2 presents a set of propositions that are proved from first principles and state common properties found in graphs, matrices, and transversals. In Sect. 2.4 these properties are combined and presented into an abstract setting, that will serve as our starting point for the definition of matroids that will be given in Chap. 3. In Chap. 3 the axiomatic definition of matroids as motivated by Sect. 2.4 is given in Sect. 3.1, and the rest of the sections in that chapter prove equivalent axiomatic definitions with the exception of Sect. 3.7 where an algorithmic definition of matroids is given. Chapter 4 contains fundamental results from matroid theory, such as representability, the notions of duality and minors, and connectivity in matroids. Chapter 5 presents a decomposition theory for graphic matroids as well as a recognition algorithm that is a result of that theory. A detailed numerical example that illustrates the theoretical results of that chapter is given in Sect. 5.4. In Chap. 6 we introduce signed-graphic matroids and present a generalization of the results given in Chap. 5. At the end of each chapter a short section with additional notes and references is provided.

Chapter 2
Graph Theory, Vector Spaces, and Transversals

In this chapter we will present a set of propositions that characterize common properties of graphs, vector spaces, and transversals. In each case emphasis has been given to derive the aforementioned properties from first principles. In the last section of the chapter the results presented in Sects. 2.1–2.3, will be unified into a common abstract framework which will serve as the starting point for the definition of matroids that will be given in Chap. 3. The purpose of this chapter is to illuminate the existence of common properties in these three seemingly different contexts.

2.1 Graph Theory

All the necessary definitions and results from graph theory will be presented in this section. The main references for this section are the books by Diestel (2006) and Wilson (1996).

By a graph $G(V, E)$ we mean a finite set of **vertices** V, and a set of **edges** which consists of 2-tuples of V. Given a graph G we will write $V(G)$ for its set of vertices and $E(G)$ for its set of edges. A graph G with $V(G) = \{v_1, v_2, v_3, v_4\}$ and $E(G) = \{e_1, e_2, e_3, e_4, e_5, e_6, e_7\}$ is shown in Fig. 2.1. Note that this graph will be referred to as a sample graph in various examples throughout the book. An edge $e = (v, v)$ is called a **loop**, while two equal edges are called **parallel edges**. A graph with no loops and parallel edges is called **simple**. Some graph H is called a **subgraph** of G and we write $H \subset G$ if $V(H) \subseteq V(G)$ and $E(H) \subseteq E(G)$. A subgraph H of G is called **proper** if $H \neq G$. The **order** of a graph G is $|V(G)|$ while its **size** is $|E(G)|$. A graph of order 1 is called **trivial**. By $G \cup H$ we mean the graph $(V(G) \cup V(H), E(G) \cup E(H))$, and by $G \cap H$ we mean the graph $(V(G) \cap V(H), E(G) \cap E(H))$. Whenever applicable the vertices that define an edge are called its **end-vertices**. For some $V' \subseteq V(G)$ the **induced** subgraph of V' in G is written as $G[V']$, and is defined by $V(G[V']) = V'$ and $E(G[V']) = \{(v, w) \in E(G) : v, w \in V'\}$. For $E' \subseteq E(G)$ the induced subgraph of E' in

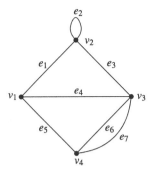

Fig. 2.1 A sample graph G

G is written as $G[E']$, and is defined as $E(G[E']) = E'$ and $V(G[E']) = \{v \in V(G) : v$ is an end-vertex of some edge in $E' \}$. The set

$$N_G(v) = \{w \in V(G) : (v, w) \in E(G)\}$$

is called the **neighborhood** of v in G, while for some $U \subseteq V(G)$ the neighborhood of U is the union of the sets $N_G(v)$ for all $v \in U$, minus U. If for two vertices $v, w \in V(G)$ we have $(v, w) \in E(G)$, then v and w are called **adjacent** vertices. If for a graph G with n vertices any two vertices are adjacent, then we say that the graph is **complete** and we write K_n. We say that an edge e is **incident** to a vertex v if $v \in e$, while if two distinct edges have a vertex in common they are **adjacent**. The **degree** $d_G(v)$ of a vertex $v \in V(G)$ is the number of edges incident to v, or equivalently $|N_G(v)|$. **Identifying** two vertices u and v is the operation where we replace u and v with a new vertex v' in both $V(G)$ and $E(G)$. The **deletion of an edge** e from G is the subgraph defined as $G \backslash \{e\} = (V(G), E(G) - \{e\})$. The subgraph of G obtained by a series of deletions of the edges $X \subseteq E(G)$ is denoted by $G \backslash X$. The **contraction of an edge** $e = (u, v)$ is the subgraph denoted by $G/\{e\}$ which results from G by identifying u, v in $G \backslash \{e\}$. The subgraph of G obtained by a series of contractions of the edges $Y \subseteq E(G)$ is denoted by G/Y. Sometimes it will be more convenient to use the complement operations of deletion and contraction. In order to make notation less cumbersome, we will write $G|X = G \backslash (E(G) - X)$ for the deletion *to* X, and $G.Y = G/(E(G) - Y)$ for contraction *to* Y. The **deletion of a vertex** v of G is defined as the deletion of all edges incident to v and the deletion of v from $V(G)$. A graph G' is called a **minor** of G if it is obtained from a sequence of deletions and contractions of edges and deletions of vertices of G. It can be easily shown that the order upon which the operations are performed is not important, so for the minor produced by the deletion of $X \subseteq E(G)$ and the contraction of $Y \subseteq E(G)$ we write $G \backslash X / Y$.

Any partition $\{T, U\}$ of $V(G)$ for nonempty T and U, defines a **cut** of G denoted by $E(T, U) \subseteq E(G)$ as the set of edges incident to a vertex in T and a vertex in U. A cut of the form $E(v, V(G) - \{v\})$ is called the **star** of vertex v. A $v_0 - v_n$ **walk** in a graph $G(V, E)$ is a subgraph of G that is defined by a sequence of vertices and edges in a consecutive manner, which starts with the vertex v_0 and ends at the vertex v_n,

$$v_0, e_1, v_1, e_2, \ldots, v_{n-1}, e_n, v_n,$$

where $e_i = (v_{i-1}, v_i)$ for $i = 1, \ldots, n$. A $s - t$ walk with all distinct vertices is called $s - t$ **path**, while if it is closed, that is, $s = t$, then it is called a **cycle**.

If there is a partition of the vertex set $V(G) = V_1 \cup \cdots \cup V_k$ such that $E(G[V_i]) = \emptyset$ for all $i = 1, \ldots, k$, then we say that G is a k-**partite** graph, where if $k = 2$ we say **bipartite**. Equivalently, we can say that in k-partite graphs we only have edges between vertices of different sets in the vertex partition. Note that any graph G is n-partite for $n = |V(G)|$. If in a k-partite graph with $V(G) = V_1 \cup \cdots \cup V_k$ and $|V_i| = n_i$ for $i = 1, \ldots, k$, we have edges between any pair of vertices in different vertex sets of the partition, then we have a complete k-partite graph, and write $K_{n_1, n_2, \ldots, n_k}$. Examples of complete graphs are shown in Fig. 2.2.

A graph that can be drawn in the plane such that no two edges intersect is called **planar**, while any such drawing of a planar graph is called a plane drawing. For example the graph in Fig. 2.1 is planar, while the graphs $K_{3,3}$ and K_5 in Fig. 2.2 are not. The continuous regions in the plane so formed by the deletion of the plane drawing of a planar graph, are called **faces**. For example in the plane drawing in Fig. 2.1, there are four faces as defined by the sets of edges $\{\{e_1, e_4, e_3\}, \{e_5, e_4, e_6\}, \{e_6, e_7\}, \{e_2\}\}$ as well as an *outer* face defined by the edges $\{e_1, e_2, e_3, e_5, e_7\}$. The **geometric dual** G^* of a planar graph G is constructed by considering any face in a plane drawing of G to be a vertex of G^* and connecting two vertices of G^* if the corresponding faces are adjacent. For example in Fig. 2.3 we draw the geometric dual of the graph in Fig. 2.1.

If we assign a direction to any edge of a graph G, we obtain a **directed** graph $\vec{G}$. If $e = (v, w) \in E(\vec{G})$ is directed from v to w, then we say that v is the **tail** and w is the **head** of e respectively. The edges in a directed graph are called **arcs**. An **orientation** of a graph $G(V, E)$ is a function that assigns to the end-vertices of each edge $e = (v, w) \in E(G)$ a sign in $\{+1, -1\}$ such that $o(e, v) = -o(e, w)$. Interpreting v as the tail of e when $o(e, v) = -1$ and the head otherwise, any orientation of a graph results in a directed graph.

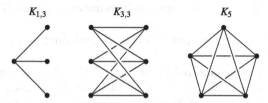

$K_{1,3}$ $\qquad$ $K_{3,3}$ $\qquad$ K_5

Fig. 2.2 Complete graphs

Fig. 2.3 The geometric dual
G^* of G in Fig. 2.1

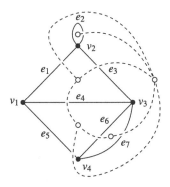

We will say that a graph is **connected** if for any $v, w \in V(G)$ there exists a $v - w$ path. One can easily show that the connectivity relationship on $V(G) \times V(G)$ is an equivalence relationship, thereby it partitions $V(G)$ into the so-called **connected components** or simply components, which are maximally connected subgraphs of G. The number of connected components of a graph G will be denoted by k_G. There are several notions of higher connectivity or k-connectivity in graphs that have appeared in the literature. The two most common ones are vertex connectivity and edge connectivity. For $k \in \mathbb{N}$ we say that a graph G is k-**vertex-connected**, if $|V(G)| > k$ and $G \backslash X$ is connected for any $X \subseteq V(G)$ with $|X| < k$. Equivalently, G is k-vertex-connected if k is the minimum number of vertices that you can delete and make G disconnected or the trivial graph K_1. We will write $\kappa(G)$ for the vertex connectivity number of a graph. For $k \in \mathbb{N}$ we say that a graph G is k-**edge-connected**, if $|E(G)| > k$ and $G \backslash Y$ is connected for any $Y \subseteq E(G)$ with $|Y| < k$. Equivalently, G is k-edge-connected if k is the minimum number of edges that you can delete and make G disconnected or the trivial graph K_1. We will write $\alpha(G)$ for the edge connectivity number of a graph. Both vertex and edge connectivities defined above use as a base notion the connectivity of a graph. Note however that while 1-vertex-connectivity is equivalent to connectivity, this is not the case for 1-edge-connectivity. Tutte introduced an alternative definition of the connectivity in a graph, in an attempt to make the connectivity of a graph and its associated graphic matroid equal, and also make the connectivity of a graph duality invariant. For $k \geq 0$, a k-**separation** of a connected graph G is a partition $\{A, B\}$ of the edges such that $\min\{|A|, |B|\} \geq k$ and $|V(G[A]) \cap V(G[B])| = k$. The **connectivity number** of a graph G is defined as

$$\lambda(G) = \min\{k : G \text{ has a } k\text{-separation}\}, \tag{2.1}$$

and we say that G is k-**connected** for any $k \leq \lambda(G)$. Thus, a k-connected graph is also l-connected for $l = 0, \ldots, k - 1$. If G does not have a k-separation for any number $k \geq 0$, then $\lambda(G) = \infty$. We could equivalently say that that a graph has a k-separation if there exist k vertices whose deletion will separate a set of k edges from another set of k edges. A **bond** in a graph G is a set of edges $Y \subseteq E(G)$ whose deletion increases the number of connected components, and is minimal with respect to that property. It is easy to see that if Y is a bond in G, then $k_{G \setminus Y} = k_G + 1$. Note that any bond is a cut, but not the other way around. Given that G is connected, the two connected components of $G \setminus Y$ will be called the **end-graphs** of Y.

The operation of **twisting** in graphs is defined as follows. Let $\{X, Y\}$ be a 2-separation of G such that $V(G[X]) \cap V(G[Y]) = \{v, u\}$. The twisting of G about v and u is the graph obtained by interchanging u and v in every edge of X. We can think of the operation as the separation of $G[X]$ and $G[Y]$ at the vertices u and v, *twisting* either subgraph about an axis perpendicular to the line that passes through u and v, and reconnecting them (see Fig. 2.4). Observe that if G' is the graph so obtained by twisting from G the cycles and bonds in both graphs remain the same. For example in Fig. 2.4 the set of edges $\{e_1, e_3, e_4, e_5\}$ is a cycle, and $\{e_3, e_5, e_6\}$ is a bond in both G and G'. However, G and G' are not isomorphic since $d_{G'}(v) = 4$ while G does not have a vertex of degree four.

A graph which contains no cycles is called a **forest** or **acyclic**, while if it is also connected it is called a **tree**. If G' is a subgraph of G and $V(G') = V(G)$ then G' is called a **spanning** subgraph of G. If a tree is a spanning subgraph of a graph, then it is called a **spanning tree**. A union of spanning trees of all connected components in a graph is called a **spanning forest**. The following is a fundamental property of trees.

Theorem 2.1 *If T is a tree, then $|E(T)| = |V(T)| - 1$.*

Proof By induction on $|V(T)|$. For $|V(T)| = 1$ we have the trivial graph with no edges. Assume that it holds for $|V(T)| < k$ and consider any tree T with $|V(T)| = k$. For some $e \in E(T)$, the graph $T \setminus e$ will consist of two trees T_1 and T_2, with $|V(T_i)| < k$ for $i = 1, 2$. By the induction hypothesis we have $|E(T_i)| = |V(T_i)| - 1$ for $i = 1, 2$. Therefore,

$$|E(T)| = |E(T_1)| + |E(T_2)| + 1 = |V(T_1)| + |V(T_2)| - 1 = |V(T)| - 1.$$

$\square$

As a corollary to Theorem 2.1 we have that $|E(T)| = |V(T)| - k_T$ for any forest T. The following result shows that spanning forests are maximally acyclic subgraphs.

Proposition 2.1 *No spanning forest of a graph G contains another spanning forest as a subgraph.*

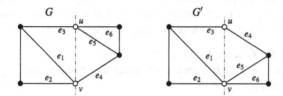

Fig. 2.4 Twisting

Proof It is enough to show that no spanning tree T of some graph G contains another spanning tree T' as a proper subgraph. Let $|V(G)| = |V(T)| = |V(T')| = n$ be the order of the graph, while for T' to be a proper subgraph of T we must have $E(T') \subset E(T)$. Therefore, $|E(T')| < |E(T)| = n - 1$ which contradicts Theorem 2.1. □

Proposition 2.2 *If X is a forest then it contains at least two vertices of degree 1.*

Proof Enough to prove the theorem for trees. Let $v_0 - v_n$ be a path in a tree T where the number of edges n is maximum. Then clearly $d_T(v_0) = d_T(v_n) = 1$, since otherwise T either contains a cycle, or a path with $n + 1$ edges. □

In the discussion that follows, we will make the assumption that the graph G under consideration is connected, unless otherwise stated. However, all the results that will be stated are directly generalizable for disconnected graphs. The following proposition is a fundamental property of acyclic graphs.

Proposition 2.3 *If X and Y are two forests of a graph G with $|E(X)| > |E(Y)|$, then there exists an edge $e \in E(X) - E(Y)$ such that $Y \cup \{e\}$ is also a forest of G.*

Proof Assume that $Y \cup \{e\}$ is not a forest for all $e \in E(X) - E(Y)$. We will show that $E(Y) \geq E(X)$. If Y is not a forest with the addition of the edge $e = (v, w)$ then the vertices v, w must belong to the same connected component of Y. This implies that $V(X) = V(Y)$ and each connected component of X is a subgraph of a connected component of Y, which means that

$$k_X \geq k_Y. \tag{2.2}$$

By Theorem 2.1 we have $|E(X)| = |V(X)| - k_X$ and $|E(Y)| = |V(Y)| - k_Y$, thus, combining with (2.2) we have

$$|E(Y)| \geq |E(X)|,$$

which is a contradiction. □

Proposition 2.4 *If T_1 and T_2 are two spanning trees of a graph G, and $e \in E(T_1) - E(T_2)$, then there exists an edge $f \in E(T_2) - E(T_1)$ such that $(T_1 \backslash \{e\}) \cup \{f\}$ is also a spanning tree of G.*

Proof Observe that $T_1 \backslash \{e\}$ is a forest of two nonempty trees, say A and B. Since T_2 is a spanning tree, it must contain an edge f distinct from e, such that $f = (v, w)$, $v \in V(A)$, $w \in V(B)$. Therefore, the subgraph $(T_1 \backslash \{e\}) \cup \{f\}$

(i)   is spanning for G since $V(A) \cup V(B) = V(G)$,
(ii) contains a $v - w$ path between any two vertices v and w, and
(iii) does not contain a cycle,

that is, a spanning tree for G. □

Proposition 2.5 *If C_1, C_2 are distinct cycles of a graph G and $e \in E(C_1) \cap E(C_2)$, then the graph $(C_1 \cup C_2)\backslash\{e\}$ contains a cycle of G.*

Proof Let $G' = (C_1 \cup C_2)\backslash\{e\}$ and $e = (v, w)$. If G' does not contain any cycle then it is a forest, and $k_{G'} \geq 1$. If the end-vertices of e reside in different connected components of G', then $C_1 \cup C_2$ does not contain a cycle, a contradiction. Therefore, both end-vertices of e are on the same connected component of G', say T, which is a tree. The graph $T \cup \{e\}$ contains a unique cycle, for otherwise there exist two distinct $v - w$ paths in T, which imply the existence of a cycle in T. This means that $G' \cup \{e\} = (C_1 \cup C_2)$ should contain a unique cycle, a contradiction. $\square$

The following result which relates cycles and bonds is easy to prove, bearing in mind that the deletion of bond from a graph always results in two components.

Proposition 2.6 *If C is a cycle and C^* a bond of G, then $|E(C) \cap E(C^*)|$ is even.*

We will now introduce the notion of the rank of a graph. Consider any graph G of order n and size m with k_G connected components. The **rank** of G is denoted by $r(G)$, and is the number of edges in any spanning forest, which in this case is clearly $r(G) = n - k_G$. The minimum number of edges which must be deleted from G to create spanning forest with k_G components is $m - n + k_G$, and we will denote this number by $\gamma(G)$.

Proposition 2.7 *For a graph G the rank satisfies the following:*

(i) $0 \leq r(G) \leq |E(G)|$,
(ii) *if H a subgraph of G then $r(H) \leq r(G)$,*
(iii) *for any pair of subgraphs H and K of G we have*

$$r(H \cup K) \leq r(H) + r(K) - r(H \cap K).$$

Proof The first two properties follow directly from the definition of the rank. We will prove the third property, which is called **submodularity** of a function. We can identify three possible cases. If $V(H) \cap V(K) = \emptyset$ then $E(H) \cap E(K) = \emptyset$, it means that H and K are disjoint subgraphs, therefore

$$r(H \cup K) = |V(H)| + |V(K)| - (k_H + k_K) = r(H) + r(K),$$

where k_H, k_K are the number of connected components of H and K respectively. If $V(H) \cap V(K) \neq \emptyset$ and $E(H) \cap E(K) = \emptyset$, then $V(H) \cap V(K)$ is a set of isolated vertices, and the result follows easily.

Consider now the case where $V(H) \cap V(K) \neq \emptyset$ and $E(H) \cap E(K) \neq \emptyset$. The subgraph $H \cap K$ of G is not empty by assumption, and let us say that T_1 is a spanning forest of $H \cap K$. Since $|E(H \cap K)| < |E(H)|$, by Proposition 2.3 T_1 can be extended into a spanning forest $T_1 \triangle T_2$. This set can be extended into a spanning forest $(T_1 \triangle T_2) \triangle T_3$ of $H \cup K$. So the subgraph defined by $T_1 \cup T_3$ must be a forest of K, but not necessarily maximal. Therefore, we have

$$r(H) + r(K) \geq |E(T_1 \cup T_2)| + |E(T_1 \cup T_3)|$$
$$= 2|E(T_1)| + |E(T_2)| + |E(T_3)|$$
$$= |E(T_1)| + |E(T_1 \cup T_2 \cup T_3)|$$
$$= r(H \cup K) + r(H \cup K).$$

$\square$

Finally, let us mention two results which relate cycles and bonds with spanning forests.

Proposition 2.8 *For any spanning forest X of a graph G,*

(i) *If C^* is a bond of a G, then $E(C^*) \cap E(X) \neq \emptyset$.*
(ii) *If C is a cycle of G, then $E(C) \cap (E(G) - E(X)) \neq \emptyset$.*

Proof (i) Let $E(C^*) \cap E(X) = \emptyset$. Then X is also a spanning forest of $G \backslash E(C^*)$ with $|V(G)| - k_G$ edges, which is a contradiction since according to Theorem 2.1 it should have $|V(G)| - k_{G \backslash E(C^*)}$ edges, where $k_{G \backslash E(C^*)} = k_G + 1$.

(ii) Since C is a subgraph of a connected component G_1 of G, we only consider the spanning tree T of G_1. If $E(C) \cap (E(G_1) - E(T)) = \emptyset$ then C is contained in T, a contradiction since trees are acyclic. $\square$

2.2 Vector Spaces

In this section we will present elementary results about finite vector spaces and linear independence. We provide as a basic reference the book by Nering (1970), but any standard textbook on linear algebra will suffice. Some familiarity is assumed with the definition of a vector space over a field $\mathbb{F}$, and the operations of vector addition and scalar multiplication. The elements of a field are called scalars and we will write x, while those of a vector space are called vectors and we will write $\mathbf{x}$.

Given a field $\mathbb{F}$ and set of m vectors $\mathbf{x}_i \in \mathbb{F}^n$ we say that $\mathbf{y}$ is a **linear combination** of the vectors $\mathbf{x}_i$ if there exist scalars $a_i \in \mathbb{F}$ such that

$$\mathbf{y} = \sum_{i=1}^{m} a_i \mathbf{x}_i.$$

A **linear relation** is an expression among the $\mathbf{x}_i$ of the form $\sum_i a_i \mathbf{x}_i = \mathbf{0}$, where $a_i \neq 0$ for some i. If $a_i = 0$ for all i then we say it is a *trivial* linear relation. A set $\{\mathbf{x}_i\}$ of vectors is said to be **linearly independent** in $\mathbb{F}$ if there does not exist a nontrivial linear relationship among them, or equivalently if $\sum_i a_i \mathbf{x}_i = \mathbf{0}$ then $a_i = 0$ for all i. If a set of vectors is not linearly independent, then we say it is **linearly dependent**.

Theorem 2.2 *A set of vectors $\{\mathbf{x}_1, \ldots, \mathbf{x}_m\}$ is linearly dependent if and only if some $\mathbf{x}_k$ is a linear combination of the $\mathbf{x}_i$ for $i < k$.*

Proof ($\Rightarrow$) Say that $\{\mathbf{x}_1, \ldots, \mathbf{x}_m\}$ is a linearly dependent set of vectors, where $\mathbf{x}_i \neq \mathbf{0}$ for all i. Then there exist scalars $a_i \in \mathbb{F}$, not all zero, such that $\sum_i a_i \mathbf{x}_i = \mathbf{0}$. Rearrange the vectors such that the corresponding scalars are

$$a_1, a_2, \ldots, a_k, a_{k+1}, \ldots, a_m,$$

and $a_j = 0$ for $j > k$. Therefore, $k \geq 2$ while

$$\mathbf{x}_k = \sum_{i=1}^{k-1} \left(-\frac{1}{a_k} a_i \right) \mathbf{x}_i,$$

and the result follows.
($\Leftarrow$) By the definition of linear independence. $\qquad\qquad\qquad\qquad\qquad\square$

Let X be the set of vectors $\{\mathbf{x}_1, \ldots, \mathbf{x}_m\} \subset \mathbb{F}^n$. The **span** of X, denoted by $\langle X \rangle$, is the set of all linear combinations of the vectors in X as

$$\langle X \rangle = \{\mathbf{x} \in \mathbb{F}^n : \mathbf{x} = \sum_i a_i \mathbf{x}_i, \mathbf{x}_i \in X, \forall a_i \in \mathbb{F}\}.$$

If for a vector space V and subspace X we have $V = \langle X \rangle$, then we say that X is a **spanning set** of V. The following are some fundamental properties of the span.

Theorem 2.3 *If $X \subseteq \langle Y \rangle$ and $Y \subseteq \langle Z \rangle$, then $X \subseteq \langle Z \rangle$.*

Proof If $Y = \{\mathbf{y}_1, \ldots, \mathbf{y}_k\}$ and $Z = \{\mathbf{z}_1, \ldots, \mathbf{z}_l\}$, then for each $\mathbf{x} \in X$ there exist scalars a_i such that

$$\mathbf{x} = \sum_{i=1}^{k} a_i \mathbf{y}_i,$$

while for each $\mathbf{y}_i \in Y$ there exist scalars b_{ij} such that

$$\mathbf{y}_i = \sum_{j=1}^{l} b_{ij} \mathbf{z}_j.$$

Therefore, we can express $\mathbf{x}$ as

$$\mathbf{x} = \sum_{i=1}^{k} a_i \left(\sum_{j=1}^{l} b_{ij} \mathbf{z}_j \right) = \sum_{j=1}^{l} \left(\sum_{i=1}^{k} a_i b_{ij} \right) \mathbf{z}_j,$$

that is, as a linear combination of the vectors in Z. $\qquad\qquad\qquad\qquad\qquad\square$

Theorem 2.4 *If $\mathbf{x}_k$ in $X = \{\mathbf{x}_1, \mathbf{x}_2, \ldots, \mathbf{x}_k\}$ is linearly dependent on $\{\mathbf{x}_1, \mathbf{x}_2, \ldots, \mathbf{x}_{k-1}\}$ then $\langle X \rangle = \langle X - \mathbf{x}_k \rangle$.*

Proof Since every vector in X can be expressed as a linear combination of the vectors in $X - \mathbf{x}_k$, we have that $X \subseteq \langle X - \mathbf{x}_k \rangle$, which in turn implies that $\langle X \rangle \subseteq \langle X - \mathbf{x}_k \rangle$. However, since $(X - \mathbf{x}_k) \subseteq X$ we have that $\langle X - \mathbf{x}_k \rangle \subseteq \langle X \rangle$. $\qquad\square$

Theorem 2.5 *For any subset X of a vector space V, $\langle X \rangle = \langle \langle X \rangle \rangle$.*

Proof Since $X \subseteq \langle X \rangle$ then $\langle X \rangle \subseteq \langle \langle X \rangle \rangle$ by Theorem 2.4. We also want to show that $\langle \langle X \rangle \rangle \subseteq \langle X \rangle$. It can be shown that for any two vector spaces Y and Z if $Y \subseteq \langle Z \rangle$ then $\langle Y \rangle \subseteq \langle Z \rangle$ also. Since $X \subseteq \langle X \rangle$ then $\langle X \rangle \subseteq \langle X \rangle$, which implies $\langle \langle X \rangle \rangle \subseteq \langle X \rangle$. $\square$

The following theorem is the so-called Steinitz Replacement Theorem, and its proof technique will be used in subsequent results.

Theorem 2.6 *If a finite set of vectors $X = \{\mathbf{x}_1, \mathbf{x}_2, \ldots, \mathbf{x}_n\}$ is a spanning set of a vector space V, then every linearly independent set in V contains at most n elements.*

Proof Consider any linearly independent set of vectors in V, say $Y = \{\mathbf{y}_1, \mathbf{y}_2, \ldots, \mathbf{y}_m\}$. We will show that $m \leq n$. The basic step in the proof would be to start replacing vectors in X by vectors in Y, while maintaining a spanning set at every step.

Suppose that at step k, the spanning set is $X_k = \{\mathbf{y}_1, \mathbf{y}_2, \ldots, \mathbf{y}_k, \mathbf{x}_{k+1}, \ldots, \mathbf{x}_n\}$. Since $\langle X_k \rangle = V$, $\mathbf{y}_{k+1}$ can be expressed as a linear combination of vectors from X_k. The main observation here is that these vectors have to include some of the $\mathbf{x}_i$'s since otherwise Y would be a linearly dependent set of vectors. Thus, we can assume that $\mathbf{x}_{k+1}$ is linearly dependent on $\{\mathbf{y}_1, \mathbf{y}_2, \ldots, \mathbf{y}_{k+1}, \mathbf{x}_{k+2}, \ldots, \mathbf{x}_n\}$ which we will denote by X_{k+1}. By Theorem 2.4 we have that $\langle X_{k+1} \rangle = \langle X_k \rangle = V$. If we continue this process and $m > n$, it would mean that $X_n = \{\mathbf{y}_1, \mathbf{y}_2, \ldots, \mathbf{y}_n\}$ is a spanning set of V, therefore $\mathbf{y}_{n+1}$ is linearly dependent on X_n, contradicting our original hypothesis that Y is a linearly independent set of vectors. $\qquad\square$

A linearly independent set of vectors which is a spanning set of a vector space V is called a **basis** of V. The linear independence of the elements in a basis B, implies the uniqueness of the coefficients in the linear combination of some vector $\mathbf{x} \in V$ expressed in terms of the vectors in B. Assume that $\mathbf{x} = \sum_{\mathbf{y}_i \in B} a_i \mathbf{y}_i$ and $\mathbf{x} = \sum_{\mathbf{y}_i \in B} b_i \mathbf{y}_i$. Then $\sum_{\mathbf{y}_i \in B} (a_i - b_i) \mathbf{y}_i = \mathbf{0}$, but since the $\mathbf{y}_i$'s are linearly independent this would mean that $a_i - b_i = 0$ for all i. The cardinality of a basis in a vector space V will be called the **dimension** of V, and we write $\dim(V)$. The next two theorems are fundamental in Linear Algebra.

Theorem 2.7 *If V is a subspace of $\mathbb{F}^n$, then*

$$dim(V) + dim(V^\perp) = n.$$

Theorem 2.8 *If V, W two subspaces then*

$$dim(V) + dim(W) = dim(V \cap W) + dim(V + W),$$

where $V + W = \{\mathbf{v} + \mathbf{w} : \mathbf{v} \in V, \mathbf{w} \in W\}$.

Given any subset X of a vector space V, we can define a function $r : V \rightarrow \mathbb{Z}_+$ which is called the **rank** of X, as the dimension of the subspace spanned by the vectors in X. In what follows we see that the dimension of a vector space is well-defined.

Theorem 2.9 *In a vector space every basis has the same number of elements.*

Proof Immediate consequence of Theorem 2.6. □

The following two theorems demonstrate that a basis can be characterized as a maximally linearly independent set or, equivalently, as a minimal spanning set.

Theorem 2.10 *Every spanning set in a vector space contains a basis.*

Proof Let X be a spanning set and $Y \subseteq X$ a maximally linearly independent subset. We know from Theorem 2.6 that $|Y| \leq |X|$. Now if $|Y| = |X|$, then Y is a basis. If $|Y| < |X|$, then by repeated application of Theorem 2.4 we can conclude that $\langle Y \rangle = \langle X \rangle$, therefore Y is a spanning set. □

Theorem 2.11 *Every linearly independent set of vectors in a vector space can be extended to a basis.*

Proof Let $A = \{\mathbf{a}_1, \ldots, \mathbf{a}_n\}$ be a basis and $B = \{\mathbf{b}_1, \ldots, \mathbf{b}_m\}$ a linearly independent set, where from Theorem 2.6 we know that $m \leq n$. If $m = n$ then there is nothing to prove. If $m < n$ consider the ordered set of vectors $C = (\mathbf{b}_1, \ldots, \mathbf{b}_m, \mathbf{a}_1, \ldots, \mathbf{a}_n)$. Clearly C is linearly dependent, so by Theorem 2.2 there exists some vector in C which can be expressed as a linear combination of the vectors preceding it. But none of the $\mathbf{b}_i$ could be such a vector since B is linearly independent, therefore it has to be one of the vectors in A, say $\mathbf{a}_i$. If we remove $\mathbf{a}_i$ from C and repeat the process on the resulting set, eventually we will end up with a linearly independent set of vectors which will contain B, and by Theorem 2.4 will span the vector space, i.e., a basis. □

We can now state the following regarding vector spaces.

Proposition 2.9 *If X and Y are linearly independent sets of vectors in V with $|X| > |Y|$, then there exists some $\mathbf{x} \in X - Y$ such that $Y \cup \{\mathbf{x}\}$ is a linearly independent set of vectors in V.*

Proof If $Y \cup \{\mathbf{x}\}$ is linearly dependent for all $\mathbf{x} \in X - Y$ then $X \subseteq \langle Y \rangle$ by Theorem 2.4, which implies that X is a linearly dependent set of vectors. □

Proposition 2.10 *If B_1 and B_2 both bases of V and $\mathbf{x} \in B_1$, then there exists some $\mathbf{y} \in B_2$ such that $(B_1 - \{\mathbf{x}\}) \cup \{\mathbf{y}\}$ is also a basis of V.*

Proof The proof can be derived from the proof of either Theorem 2.6 or Theorem 2.11. □

Proposition 2.11 *If X and Y are two distinct minimally linearly dependent sets of vectors in V and $\mathbf{z} \in X \cap Y$, then the set $(X \cup Y) - \{\mathbf{z}\}$ contains a minimally linearly dependent set of vectors in V.*

Proof It is enough to show that the set of vectors $(X \cup Y) - \{z\}$ is linearly dependent. Let $X = \{x_1, \ldots, x_k, z_1, \ldots, z_{t+1}\}$, $Y = \{z_1, \ldots, z_{t+1}, y_1, \ldots, y_s\}$, and assume that $z = z_{t+1}$. Since both X and Y are minimally linearly dependent, we can express z as a linear combination of the vectors in $X - \{z\}$ and $Y - \{z\}$, such that in both cases all scalars used in the linear combination are not zero. Thus, there exist scalars a_i, b_i, c_i and d_i, all nonzero, such that

$$z = a_1 x_1 + \cdots + a_k x_k + b_1 z_1 + \cdots + b_t z_t, \tag{2.3}$$

$$z = c_1 x_1 + \cdots + c_k x_k + d_1 y_1 + \cdots + d_s y_s. \tag{2.4}$$

Subtracting (2.4) from (2.3) we get

$$a_1 x_1 + \cdots + a_k x_k + (b_1 - c_1) z_1 + \cdots + (b_t - c_t) z_t - d_1 y_1 - \cdots - d_s y_s = \mathbf{0},$$

where all the scalars are not zero, therefore $(X \cup Y) - \{z\}$ is linearly dependent. $\square$

Finally, the next proposition can be proved directly from the definition of the rank and Theorem 2.8.

Proposition 2.12 *The rank function $r : 2^V \to \mathbb{Z}_+$ satisfies:*

(i) $0 \leq r(X) \leq |X|$, *for any $X \subseteq V$.*
(ii) *if $X \subseteq Y$ then $r(X) \leq r(Y)$.*
(iii) *for any X, Y we have $r(X \cup Y) + r(X \cap Y) \leq r(X) + r(Y)$ (submodularity).*

2.3 Transversal Theory

Here we will present fundamental results in finite Transversal Theory. The main references for this section are the works of Mirsky (1969, 1971).

Given some ground set E and a family $\mathscr{F} = (S_i : i \in I)$ of subsets of E, a **transversal** of $\mathscr{F}$ is set $X = \{x_1, x_2, \ldots, x_{|I|}\} \subseteq E$ of distinct elements such that $x_i \in S_i$ $\forall i \in I$. Given E and $\mathscr{F}$ as previously, a transversal of some subfamily $(S_i : i \in J \subset I)$ is called a **partial transversal** of $\mathscr{F}$. So given a family of n subsets of a ground set E, a transversal is a set of n distinct elements from E, such that each subset contains at least one of these elements or is represented by at least one element in the transversal. Transversals are also known in the literature as *systems of distinct representatives*. Observe that for a given transversal $X = \{x_1, x_2, \ldots, x_n\}$ of $(S_i : i \in I)$ we have no information regarding the membership of its elements with respect to the subsets S_i. If we wish to specify membership, we can *index* the transversal according to I and obtain a family of elements of X, denoted by $X = \{x_i\}_{i \in I}$, where $x_i \in S_i$ for all $i \in I$. Note that a transversal may be indexed in more than one way, while an indexed transversal corresponds to a unique transversal.

Fig. 2.5 Bipartite graph of a
set system

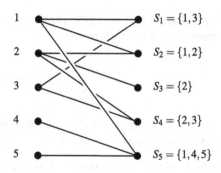

Partial transversals can also be indexed to specify membership to the subsets of the corresponding subfamily of subsets.

Example 2.1 Given $E = \{1, 2, 3, 4, 5\}$ consider the subsets

$$S_1 = \{1, 3\},$$
$$S_2 = \{1, 2\},$$
$$S_3 = \{2\},$$
$$S_4 = \{2, 3\},$$
$$S_5 = \{1, 4, 5\},$$

and let the index set be $I = \{1, 2, 3, 4, 5\}$. Then $\mathscr{F} = (S_i : i \in I)$ does not have a transversal, since we cannot find five distinct elements of E each from a different subset S_i. But $\mathscr{F}$ has a partial transversal, say $X = \{1, 2, 3, 5\}$, which is a transversal of the family of subsets $(S_i : i \in I - \{3\})$. We can also index the partial transversal and write $\{1_1, 2_2, 3_4, 5_5\}$ or alternatively $\{3_1, 1_2, 2_4, 5_5\}$.

An equivalent way of graphically representing transversals is through bipartite graphs. Given a set system $(E, \mathscr{F})$ where $\mathscr{F} = (S_i : i \in I)$, construct the bipartite graph $G(V_1 \cup V_2, E_G)$ where $V_1 = E$, $V_2 = \mathscr{F}$ and $E_G = \{(i, j) : i \in E$ and $i \in S_j\}$. So the edges of the bipartite graph represent the membership of the elements in E to the subsets in $\mathscr{F}$. The bipartite graph corresponding to the set system of Example 2.1 is shown in Fig. 2.5. A **matching** $M \subseteq E(G)$ in a graph G is a set of non-adjacent edges, while we say that M is a matching of some $U \subseteq V(G)$ if every vertex in U is an end-vertex of an edge in M. We can see that any matching in a bipartite graph of a set system corresponds to a partial transversal, and any matching of V_2 corresponds to a transversal.

Transversal theory is concerned with the question of existence of transversals or partial transversals in a set system, under some constraints. Examining the definition of transversal, it is trivial to see that if the family $(S_i : i \in I)$ of subsets of E has a transversal, then the union of any k subsets from $\mathscr{F}$ will contain at least k elements. Hall's classical theorem also establishes the sufficiency of this condition.

Theorem 2.12 (Hall 1935) *The family $\mathscr{F} = (S_i : i \in I)$ of subsets of E has a transversal if and only if*

$$\left| \bigcup_{i \in J} S_i \right| \geq |J|, \quad \text{for each} \quad J \subseteq I. \tag{2.5}$$

Proof Necessity follows from the definition of transversals. For sufficiency we will first prove the following claim:

Claim If Hall's condition in (2.5) is true for $(S_i : i \in I)$ and one of the subsets, say S_1, contains more than one element, then deleting any element from S_1 will not make the condition false for the resulting family of subsets.

Assume by contradiction that there exist $x, y \in S_1$ such that the deletion of any one of these elements from S_1 will make condition (2.5) false for the resulting family of subsets. Therefore, there will be $I_x, I_y \subseteq I - \{1\}$, such that for

$$P = \left(\bigcup_{i \in I_x} S_i \right) \cup S_1 - \{x\} \quad \text{and} \quad Q = \left(\bigcup_{i \in I_y} S_i \right) \cup S_1 - \{y\},$$

we will have

$$|P| < |I_x| + 1, \quad \text{and} \quad |Q| < |I_y| + 1. \tag{2.6}$$

It follows that

$$|P \cup Q| = \left| \left(\bigcup_{i \in I_x \cup I_y} S_i \right) \cup S_1 \right|, \tag{2.7}$$

$$|P \cap Q| = \left| \left(\bigcup_{i \in I_x \cap I_y} S_i \right) \cup S_1 - \{x, y\} \right| \geq \left| \bigcup_{i \in I_x \cap I_y} S_i \right|. \tag{2.8}$$

Combining (2.6)–(2.8) we have

$$\begin{aligned}
|I_x| + |I_y| &\geq |P| + |Q| \\
&= |P \cup Q| + |P \cap Q| \\
&\geq \left| \left(\bigcup_{i \in I_x \cup I_y} S_i \right) \cup S_1 \right| + \left| \bigcup_{i \in I_x \cap I_y} S_i \right| \\
&\geq |I_x \cup I_y| + 1 + |I_x \cap I_y| \\
&= |I_x| + |I_y| + 1,
\end{aligned}$$

which is a contradiction. We thus have that the deletion of any $x \in S_1$ will not affect the validity of Hall's condition (2.5), and the claim is proved.

Given now a family $\mathscr{F} = (S_i : i \in I)$ which satisfies (2.5), by a series of deletions we can create a family $(S_i' : i \in I)$ which also satisfies (2.5) and $|S_i'| = 1$ for all $i \in I$, which is a transversal of $\mathscr{F}$. $\qquad\square$

Let us look at partial transversals now. Given a set system $(E, \mathscr{F})$ if $\mathscr{F}$ has a transversal it obviously has a partial transversal, but the existence of a partial transversal does not necessarily imply the existence of a transversal of $\mathscr{F}$. The following theorem by Ore generalizes Hall's theorem, by providing a condition for the existence of a partial transversal of given size.

Theorem 2.13 (Ore 1955) *The family* $\mathscr{F} = (S_i : i \in I)$ *of subsets of E has a partial transversal of size k if and only if*

$$\left| \bigcup_{i \in J} S_i \right| \geq |J| + k - |I|, \quad \text{for each } J \subseteq I. \tag{2.9}$$

Proof For $k = |I|$ we have Hall's theorem. Let D be a subset of E, such that $D \neq \emptyset$ and $|D| = |I| - k$. Consider now the family of subsets

$$\mathscr{F}^* = \{S_1 \cup D, S_2 \cup D, \ldots, S_{|I|} \cup D\}.$$

Observe that $\mathscr{F}$ has a partial transversal of size k if and only if $\mathscr{F}^*$ has a transversal. This is so, since if we assume that T is a transversal of $\mathscr{F}^*$, then T can have at most $|I| - k$ distinct elements from D which implies that it has at least k elements from the subsets S_i, $i \in I$. Therefore, $\mathscr{F}$ has a partial transversal of size at least k. To show the other direction we use the same argument.

Now for $\mathscr{F}^*$ to have a transversal it must satisfy (2.5), that is for any $J \subseteq I$

$$\left| \bigcup_{i \in J} (S_i \cup D) \right| \geq |J| \quad \Leftrightarrow$$

$$\left| \bigcup_{i \in J} S_i \right| + |D| \geq |J| \quad \Leftrightarrow$$

$$\left| \bigcup_{i \in J} S_i \right| \geq |J| + k - |I|.$$

$\qquad\square$

For some family $(S_i : i \in I)$ and X a transversal of $(S_i : i \in I_X)$ for some $I_X \subseteq I$, we say that S_j is *represented* by X if $j \in I_X$.

Proposition 2.13 *If X and Y partial transversals of $\mathscr{F} = (S_i : i \in I)$ with $|X| > |Y|$, then there exists some $x \in X - Y$ such that $Y \cup \{x\}$ is a partial transversal of $\mathscr{F}$.*

Proof Since X and Y are partial transversals of $\mathscr{F} = (S_i : i \in I)$ there exist $I_X, I_Y \subseteq I$ with $|I_X| > |I_Y|$, such that X and Y are transversals of $(S_i : i \in I_X)$ and $(S_i : i \in I_Y)$ respectively. We will show that for any $x \in X - Y$ there exists $I' \subseteq I_X \cup I_Y$ such that $Y \cup \{x\}$ is a transversal of $(S_i : i \in I')$, i.e., a partial transversal of $\mathscr{F}$.

Apply an arbitrary indexing to the elements of X and Y and obtain the two families $X = \{x_i\}_{i \in I_X}$ and $Y = \{y_i\}_{i \in I_Y}$. Consider any $x_k \in X$ where $k \in I_X - I_Y$, which implies that S_k is not represented by Y. If $x_k \neq y_i$ for all $i \in I_Y$, then the result follows for $x = x_k$. If $x_k = y_i$ for some $i \in I_Y$, then $x_k \neq y_j$ for all $j \in I_Y$ different from i. Moreover $y_i \in S_k$ and $x_k \in S_i$, which means that $Y \cup \{x_k\}$ is a transversal of $(S_i : i \in I_Y \cup \{k\})$.   □

An immediate consequence of Proposition 2.13 is that maximal partial transversals, that is, partial transversals which are not properly contained in any other partial transversal, are simply transversals. As the following proposition demonstrates, transversals also obey an exchange type property, which provides a linkage for all the transversals of a family of subsets.

Proposition 2.14 *If X and Y two transversals of $\mathscr{F} = (S_i : i \in I)$ and $x \in X - Y$, then there exists $y \in Y - X$ such that $(X - \{x\}) \cup \{y\}$ is a transversal of $\mathscr{F}$.*

Proof Let $|I| = n$ while $X = \{x_i\}_{i \in I}$ and $Y = \{y_i\}_{i \in I}$ be two arbitrary indexings of the transversals X and Y respectively. Without loss of generality let $x = x_1$, and define $X_1 = X - Y$, $X_2 = X - X_1$ and $Y_1 = Y - X$, $Y_2 = Y - Y_1$. Since the elements of X and Y are distinct, we will have that $X_2 = Y_2$, and assume that $X_2 = \{x_{k+1}, \ldots, x_n\}$. Rearranging the elements of Y such that the last $n - k$ elements will be Y_2 we can write the two sets as

$$X = \{\overbrace{x_1, x_2, \ldots, x_k,}^{X_1} \overbrace{x_{k+1}, \ldots, x_{n-1}, x_n}^{X_2}\},$$
$$Y = \{\underbrace{y_{i_1}, y_{i_2}, \ldots, y_{i_k},}_{Y_1} \underbrace{y_{i_{k+1}}, \ldots, y_{i_{n-1}}, y_{i_n}}_{Y_2}\},$$

where $x_i = y_{i_j}$ for $j = k + 1, \ldots, n$, and $|X_1| = |Y_1| \geq 1$ by assumption.

Note that y_1 would be the element of Y that represents S_1. If $y_1 \in Y_1$, then $(X - \{x_1\}) \cup \{y_1\}$ is a transversal of $\mathscr{F}$, since $y_1 \neq x_l$ for $l = 2, \ldots, n$ and all the subsets S_i are represented. So let $y_1 \in Y_2$, and assume that $i_n = 1$, that is, its unique equal element in X_2 is x_n. So we will have that $x_n = y_1$, therefore $\{x_n, x_1\} \subseteq S_1$, which means that an alternative indexing exists for X, where the values of x_1 and x_n are interchanged. Let us consider the element y_n now. If $y_n \in Y_1$, then $(X - \{x_1\}) \cup \{y_n\}$ is a transversal of $\mathscr{F}$, since S_1 is represented by x_n, S_n is represented by y_n, and $y_n \neq x_l$ for $l = 2, \ldots, n$. If $y_n \in Y_2$, then we can assume that

$i_{n-1} = n$, which implies that $x_{n-1} = y_n$ therefore $\{x_{n-1}, x_n\} \subseteq S_n$. As previously, we now consider y_{n-1} and if it is contained in Y_1 then $(X - \{x_1\}) \cup \{y_{n-1}\}$ is a transversal of $\mathscr{F}$, since S_1 is represented by x_n, S_n by x_{n-1} and S_{n-1} by y_{n-1}, while all the elements are distinct. Since $Y_1 \neq \emptyset$, if we continue this process eventually we will have $y_k \in Y_1$ for some $k \in \{n, n-1, \ldots, k+1\}$, and $(X - \{x_1\}) \cup \{y_k\}$ will be a transversal of $\mathscr{F}$. $\square$

We will consider now sets $X \subseteq E$, which are not partial transversals. A **circuit transversal** of some family of subsets $\mathscr{F} = (S_i : i \in I)$ of E, is any $X \subseteq E$ which is not a partial transversal, and $X - \{e\}$ is a partial transversal for any $e \in X$. So, circuit transversals are subsets of E which are not partial transversals and are minimal with respect to that property. It follows from the definition that an $X \subseteq E$ is not a partial transversal if and only if it contains a circuit transversal. In fact we can make a stronger statement, similar to the exchange property of maximal partial transversals given in Proposition 2.14.

Proposition 2.15 *If X, Y circuit transversals of $\mathscr{F} = (S_i : i \in I)$, $X \neq Y$ and $e \in X \cap Y$, then the set $(X \cup Y) - \{e\}$ contains a circuit transversal.*

Proof It is enough to show that $C = (X \cup Y) - \{e\}$ is not a partial transversal. Assume the contrary, and let an arbitrary indexing of C be the following:

$$C = \{x_{i_1}, x_{i_2}, \ldots, x_{i_n}, y_{j_1}, y_{j_2}, \ldots, y_{j_m}\},$$

where for $I_X = \{i_1, \ldots, i_n\}$ and $I_Y = \{j_1, \ldots, j_m\}$ we have that $X = \{x_{i_k}\}_{k \in I_X} \cup \{e\}$ and $Y = \{y_{j_k}\}_{k \in I_Y} \cup \{e\}$. So C is a transversal of $(S_j : j \in I_X \cup I_Y)$. If $e \in S_k$ for some $k \in I_Y$ or $k \in I_X$, then X or Y is a transversal of $(S_j : j \in I_X \cup \{k\})$ or $(S_j : j \in I_Y \cup \{k\})$ respectively. If $e \in S_k$ for some $k \in I - (I_X \cup I_Y)$ then both X and Y are transversals of $(S_j : j \in I_X \cup I_Y \cup \{k\})$. We have that in all cases either X or Y, or both, are partial transversal of $\mathscr{F}$, a contradiction. $\square$

For a set system $(E, \mathscr{F})$ define the **rank** of a set $X \subseteq E$ as the size of the maximum partial transversal contained in X. We will write $r(X)$.

Proposition 2.16 *Given a ground set E and a family $\mathscr{F} = (S_i : i \in I)$ of subsets of E, the rank function $r : 2^E \to \mathbb{Z}_+$ satisfies:*

(i) $0 \leq r(X) \leq |X|$, *for any $X \subseteq E$;*
(ii) *if $X \subseteq Y$ then $r(X) \leq r(Y)$;*
(iii) *for any X, Y we have $r(X \cup Y) + r(X \cap Y) \leq r(X) + r(Y)$.*

Proof Properties (i) and (ii) follow from the definition of partial transversals. For (iii) define

$$I_X = \{i \in I : S_i \cap X \neq \emptyset\}, \quad I_Y = \{i \in I : S_i \cap Y \neq \emptyset\},$$

and $I_{X \cup Y} = I_X \cup I_Y$, $I_{X \cap Y} = I_X \cap I_Y$. If $I_{X \cap Y} = \emptyset$ then $X \cap Y = \emptyset$ which implies that $r(X \cap Y) = 0$ and property (iii) holds with equality. So assume that

$I_{X \cap Y} \neq \emptyset$. Consider any $J \subseteq I_{X \cup Y}$, which will also be a subset of I_X, I_Y, and $I_{X \cup Y}$. By the definition of the rank and Theorem 2.13 we know that if $\mathscr{F}$ contains a partial transversal of size $r(X)$ then

$$\left| \bigcup_{i \in J} S_i \right| \geq |J| + r(X) - |I_X|. \tag{2.10}$$

Similarly for $r(Y)$, $r(X \cup Y)$, and $r(X \cap Y)$ we have

$$\left| \bigcup_{i \in J} S_i \right| \geq |J| + r(Y) - |I_Y|, \tag{2.11}$$

$$\left| \bigcup_{i \in J} S_i \right| \geq |J| + r(X \cup Y) - |I_{X \cup Y}|, \tag{2.12}$$

$$\left| \bigcup_{i \in J} S_i \right| \geq |J| + r(X \cap Y) - |I_{X \cap Y}|. \tag{2.13}$$

Subtracting (2.10) and (2.11) from the addition of (2.12) and (2.13) we get

$$r(X) + r(Y) \geq r(X \cup Y) + r(X \cap Y) - |I_{X \cup Y}| - |I_{X \cap Y}| + |I_X| + |I_Y|$$
$$= r(X \cup Y) + r(X \cap Y).$$

$\square$

2.4 Abstract Independence

We are now in a position to raise the level of abstraction with respect to the one upon which graphs, vector spaces, and transversals reside. It should be clear to the reader from the results in Sects. 2.1–2.3, that certain sets of edges in graphs, vectors in vector spaces, and elements in transversals exhibit common behavior. Specifically, we can identify the following:

- Forests, linearly independent sets of vectors, and partial transversals have the *augmentation* property as demonstrated in Propositions 2.3, 2.9, and 2.13.
- Spanning trees, bases, and transversals have the *exchange* property as demonstrated in Propositions 2.4, 2.10, and 2.14.
- Cycles, minimally linearly dependent sets of vectors, and circuit transversals have the *elimination* property as demonstrated in Propositions 2.5, 2.11, and 2.15.
- The notion of rank in graphs, vectors, and transversals obeys the *submodularity* property in all three systems, as demonstrated in Propositions 2.7, 2.12, and 2.16.

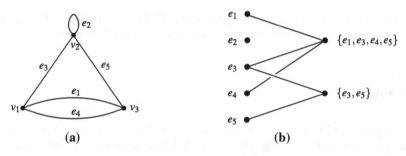

Fig. 2.6 A graph and a transversal. (a) Graph $G(V, E)$. (b) Bipartite graph of $(E, \mathscr{F})$

Therefore, we could argue that the above propositions do not describe properties that graphs, vectors, or transversals have, but rather properties of a more general and abstract notion that these systems follow. Consider a set $E = \{e_1, e_2, e_3, e_4, e_5\}$, and let us impose a structure on the elements of 2^E as a consequence of letting E be:

– the set of edges of the graph depicted in Fig. 2.6a,
– the ground set of the set system $(E, \mathscr{F})$ with $\mathscr{F} = \{\{e_1, e_3, e_4, e_5\}, \{e_3, e_5\}\}$ as illustrated by the bipartite graph in Fig. 2.6b,
– the set of columns of the matrix

$$\begin{array}{cccc} & e_1 \ e_2 \ e_3 \ \ e_4 \ e_5 \end{array}$$
$$A_E = \begin{bmatrix} -1 & 0 & -1 & -1 & 0 \\ 2 & 0 & 1 & 2 & 1 \end{bmatrix}.$$

The following subsets of 2^E are of interest to us:

$$\mathscr{B} = \{\{e_1, e_3\}, \{e_1, e_5\}, \{e_3, e_4\}, \{e_4, e_5\}, \{e_3, e_5\}\},$$
$$\mathscr{I} = \{\emptyset, \{e_1\}, \{e_3\}, \{e_4\}, \{e_5\}\} \cup \mathscr{B},$$
$$\mathscr{C} = \{\{e_2\}, \{e_1, e_4\}, \{e_3, e_4, e_5\}, \{e_1, e_3, e_5\}\}.$$

We can observe that for the graph G, set system $(E, \mathscr{F})$ and column space of A_E, the above families of subsets correspond to:

– $\mathscr{B}$ is the set of spanning trees, transversals, and bases,
– $\mathscr{I}$ is the set of forests, partial transversals, and linearly independent columns,
– $\mathscr{C}$ is the set of cycles, circuit transversals, and minimally linearly dependent columns.

Moreover, we can see that given one family we can define the other two. For example if we had $\mathscr{B}$ we have

$$\mathscr{I} = \{X \subseteq E : \exists B \in \mathscr{B} \text{ such that } X \subseteq B\}$$
$$\mathscr{C} = \{X \subseteq E : X \in 2^E - \mathscr{I}, \ X \text{ is sminimal}\}.$$

In Chap. 3 we will show that the set systems $(E, \mathcal{B})$, $(E, \mathcal{I})$, and $(E, \mathcal{C})$ are alternative representations of the same object, which will be called a matroid.

2.5 Notes

Various proofs of Theorem 2.12 which are simpler than Hall's original proof have appeared in the literature. The proof presented here is by Rado (1967), and the reduction technique used in the argument can also be applied to prove generalizations of Hall's theorem, as demonstrated by Welsh (1971). The proof of Theorem 2.13 is by Mirsky (1969).

Chapter 3
Definition of Matroids

In this chapter the common properties of graphs, vectors, and transversals presented in Chap. 2 will be considered as different sets of axioms and it will be proven that they define the same object. Several equivalent axiomatic definitions of matroids will be given, along with an algorithmic definition, which is the fundamental link between matroids and combinatorial optimization.

3.1 Independent Sets

Definition 3.1 (**Independence Axioms**) Given some finite set E, the set system $(E, \mathscr{I})$ is a **matroid** if the following are satisfied:

(I1) $\emptyset \in \mathscr{I}$.
(I2) If $X \in \mathscr{I}$ and $Y \subseteq X$ then $Y \in \mathscr{I}$.
(I3) If $X, Y \in \mathscr{I}$ and $|X| > |Y|$ then there exists $x \in X - Y$ such that $Y \cup \{x\} \in \mathscr{I}$.

We write $M(E, \mathscr{I})$ or simply M if E and $\mathscr{I}$ are self-evident.

Axiom (I3) will be called the *independence augmentation axiom* and it generalizes Propositions 2.3, 2.9, and 2.13. Any set system $(E, \mathscr{I})$ with $\mathscr{I}$ satisfying axioms (I1) and (I2) will be called an *independence system*. Since any matroid is also an independence system by definition, any future definition on independence systems applies to matroids as well. Consider an independence system $(E, \mathscr{I})$. The members of $\mathscr{I}$ are called *independent* while those of $2^E - \mathscr{I}$ *dependent*. The collection of independent sets for some $X \subseteq E$ will be

$$\mathscr{I}(X) = \{Y \subseteq X : Y \in \mathscr{I}\}.$$

Example 3.1 Consider for example the graph in Fig. 2.1 and let the ground set be $E = \{e_1, e_2, \ldots, e_7\}$. If

$$\mathscr{I}_1 = \{X \subseteq E : G[X] \text{ does not contain any cycle}\},$$

L. S. Pitsoulis, *Topics in Matroid Theory*,
SpringerBriefs in Optimization, DOI: 10.1007/978-1-4614-8957-3_3,
© Leonidas S. Pitsoulis 2014

then the set system $(E, \mathscr{I}_1)$ trivially satisfies axioms (I1) and (I2), while (I3) is Proposition 2.3. Therefore, $(E, \mathscr{I}_1)$ is a matroid. Let now

$$\mathscr{I}_2 = \{X \subseteq E : X \text{ is a matching in } G\}.$$

One can check that $(E, \mathscr{I}_2)$ is an independence system since axioms (I1) and (I2) are satisfied. However axiom (I3) is not satisfied, since if we take $X = \{e_1, e_6\}$ and $Y = \{e_4\}$ we have that both $Y \cup \{e_1\}$ and $Y \cup \{e_6\}$ are not in $\mathscr{I}_2$.   □

The augmentation axiom can be further strengthened, as follows.

Theorem 3.1 *Suppose that X, Y are independent sets in a matroid $M(E, \mathscr{I})$ and $|X| > |Y|$. Then there exists some $Z \subseteq X - Y$ such that $|Y \cup Z| = |X|$ and $Y \cup Z \in \mathscr{I}$.*

Proof Let $Z \subseteq X - Y$ be a maximal set such that $Y \cup Z \in \mathscr{I}$ and assume that $|Y \cup Z| < |X|$. We know from (I3) that such a Z exists, at least for $|Z| = 1$. Since both X and $Y \cup Z$ are independent, there exists some $x \in X - (Y \cup Z)$ such that $(Y \cup Z) \cup \{x\} \in \mathscr{I}$. Since $x \notin Z$ it implies that Z is not maximal, a contradiction. □

Two matroids M_1 and M_2 are *isomorphic* and we write $M_1 \cong M_2$, if there exists a bijection $\phi : E(M_1) \to E(M_2)$ such that $X \in \mathscr{I}(M_1)$ if and only if $\phi(X) \in \mathscr{I}(M_2)$ for all $X \subseteq E(M_1)$. We are now in a position to define the three major classes of matroids that arise from graphs, matrices, and transversals. In the definitions that follow, the fact that the mentioned independent systems are matroids is proved in Propositions 2.3, 2.9, and 2.13.

Definition 3.2 (Graphic Matroids) A matroid isomorphic to the matroid $M(G)$ with ground set $E = E(G)$ and independence family

$$\mathscr{I} = \{X \subseteq E : G[X] \text{ is a forest}\},$$

for a graph G, will be called **graphic matroid**.

Definition 3.3 (Representable Matroids) A matroid isomorphic to the matroid $M[A]$ with ground set $E = \{\text{set of columns of } A\}$ and independence family

$$\mathscr{I} = \{X \subseteq E : X \text{ is a linearly independent set of vectors in } \mathbb{F}\},$$

for a matrix $A \in \mathbb{F}^{m \times n}$ in some field $\mathbb{F}$, will be called $\mathbb{F}$-**representable matroid** or simply **representable**.

Definition 3.4 (Transversal Matroids) A matroid isomorphic to the matroid $M(E, \mathscr{F})$ with independence family

$$\mathscr{I} = \{X \subseteq E : X \text{ is a partial transversal of } \mathscr{F}\},$$

for a set system $(E, \mathscr{F})$, will be called **transversal matroid**.

For a given a graph G the matroid $M(G)$ will be called the *cycle* matroid of G, while for a given matrix A the matroid $M[A]$ will be called the *vector* matroid of A. Moreover, $GF(2)$-representable matroids will be called *binary*.

3.2 Bases

Definition 3.5 (Bases) Given an independence system $(E, \mathscr{I})$, the maximal independent sets will be called **bases**. The family of bases will be denoted by $\mathscr{B}$.

The collection of bases for some $X \subseteq E$ will be denoted by $\mathscr{B}(X)$ and is defined as

$$\mathscr{B}(X) = \{Y \subseteq X : Y \in \mathscr{I}, Y \cup \{x\} \notin \mathscr{I} \text{ for all } x \in X - Y\}. \quad (3.1)$$

It follows that $\mathscr{B}(E) = \mathscr{B}$. Note that bases in independence systems can have different cardinalities. Consider for example $X = \{e_1, e_4, e_6\}$ for the independence system $(E, \mathscr{I}_2)$ given in Example 3.1, where we have that the two maximal matchings contained in X are

$$\mathscr{B}(X) = \{\{e_1, e_6\}, \{e_4\}\}.$$

As the next important lemma shows, equal cardinality of bases is a necessary and sufficient condition for an independence system to be a matroid.

Lemma 3.1 *An independence system $(E, \mathscr{I})$ is a matroid if an only if for any $X \subseteq E$ all bases of X have the same cardinality.*

Proof Let $(E, \mathscr{I})$ be a matroid and consider some $X \subset E$. Assume by contradiction that there exist $B_1, B_2 \in \mathscr{B}(X)$ with $|B_1| > |B_2|$. Since $B_1, B_2 \in \mathscr{I}$ by axiom (I3) we can find some $x \in B_1 - B_2$ such that $B_2 \cup \{x\} \in \mathscr{I}$. Thus, B_2 is not maximally independent, a contradiction.

Let now $(E, \mathscr{I})$ be an independence system, $X, Y \in \mathscr{I}$ with $|X| > |Y|$, and consider the set $X \cup Y$. By assumption all bases of $X \cup Y$ have the same size, which is at least $|X|$ since $X \in \mathscr{I}$. This implies that Y is not a base of $X \cup Y$, so there exists some $x \in (X \cup Y) - Y = X - Y$ such that $Y \cup \{x\} \in \mathscr{I}$, which is axiom (I3) of Definition 3.1. $\square$

Theorem 3.2 (Basis Axioms) *A collection $\mathscr{B} \subseteq 2^E$ is the set of bases of a matroid $M(E, \mathscr{I})$ if and only if the following are satisfied:*

(B1) $\mathscr{B} \neq \emptyset$.
(B2) *If $B_1, B_2 \in \mathscr{B}$ and $x \in B_1 - B_2$ then there exists $y \in B_2 - B_1$ such that $(B_1 - \{x\}) \cup \{y\} \in \mathscr{B}$.*

Proof Let $\mathscr{B}$ be the set of bases of the matroid $M(E, \mathscr{I})$, where $\mathscr{I}$ satisfies axioms (I1)–(I3) of Definition 3.1. We will show that $\mathscr{B}$ satisfies (B1) and (B2). Since by axiom (I1) $\mathscr{I}$ contains at least one element, the empty set, we have that $\mathscr{B} \neq \emptyset$. Consider now $B_1, B_2 \in \mathscr{B}$, and $x \in B_1 - B_2$. Then $B_1 - \{x\} \in \mathscr{I}$ by axiom (I2) and by axiom (I3) there exists some $y \in B_2 - (B_1 - \{x\})$ such that $B = (B_1 - \{x\}) \cup \{y\} \in \mathscr{I}$. By Lemma 3.1 we have that $B \in \mathscr{B}$ since all bases of a matroid have the same cardinality and (B2) is proved.

Let $\mathscr{B} \subseteq 2^E$ satisfy (B1) and (B2). We will show that for

$$\mathscr{I} = \{X \subseteq E : \text{ there exists } B \in \mathscr{B} \text{ such that } X \subseteq B\} \tag{3.2}$$

the set system $(E, \mathscr{I})$ satisfies axioms (I1)–(I3) of Definition 3.1. Axioms (I1) and (I2) hold trivially.

To show (I3) consider any two $X, Y \in \mathscr{I}$ with $|X| > |Y|$. We have to show that there exists $x \in X - Y$ such that $Y \cup \{x\} \in \mathscr{I}$. Since $X, Y \in \mathscr{I}$, by (3.2) there exist $B_1, B_2 \in \mathscr{B}$ such that $X \subseteq B_1$ and $Y \subseteq B_2$. From all the possible B_1, B_2 such pairs, choose the one such that $|B_1 \cap B_2|$ is maximum. If $B_2 \cap (X - Y) \neq \emptyset$ then we are done, since for any element x from this set we have $Y \cup \{x\} \subseteq B_2$, so the set $Y \cup \{x\}$ is independent. Assume now by contradiction that $B_2 \cap (X - Y) = \emptyset$. If $|B_1| > |B_2|$, then by repeated application of $(B2)$ we can create a set $B = (B_1 \cap B_2) \cup S$ for some $S \subseteq B_1 - B_2$, such that $B \in \mathscr{B}$ and $B \subseteq B_1$. This contradicts the maximality of $|B_1 \cap B_2|$ and we must have $|B_1| = |B_2|$. We will partition B_1 and B_2 as illustrated in Fig. 3.1, by ordering subsets of their elements as follows. Let X and Y in B_1 and B_2 respectively, appear first. Shaded areas indicate common subsets of elements between B_1 and B_2. Partition now B_2, by first indicating the common subsets $X \cup Y$, $B_2 \cap (X - Y)$ and $(B_1 - X) \cap (B_2 - Y)$. From Fig. 3.1 it is now evident that if $B_2 \cap (X - Y) = \emptyset$ we will have

$$\begin{aligned} |B_2| &= |X \cap Y| + |Y \cap (B_1 - X)| + |(B_1 - X) \cap (B_2 - Y)| + |Y - B_1| \\ &\quad + |(B_2 - B_1) - Y| \\ &= |B_1 \cap B_2| + |Y - B_1| + |(B_2 - B_1) - Y| \\ &= |B_1|. \end{aligned} \tag{3.3}$$

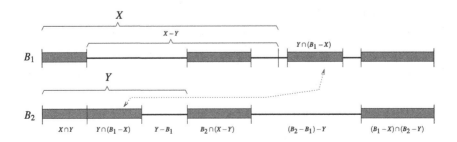

Fig. 3.1 The elements of B_1 and B_2

From Fig. 3.1 we can also see that if $B_2 \cap (X - Y) = \emptyset$ then

$$|B_1| \geq |B_1 \cap B_2| + |X - Y|. \tag{3.4}$$

Combining (3.3) and (3.4) we get

$$|Y - B_1| + |(B_2 - B_1) - Y| \geq |X - Y|. \tag{3.5}$$

Since $|X| > |Y|$ we have that

$$|X - Y| > |Y - X| \geq |Y - B_1|$$

which from (3.5) implies that $(B_2 - B_1) - Y \neq \emptyset$. So for some $y \in (B_2 - B_1) - Y$ by axiom (B2) there exists $x \in B_1 - B_2$ such that $(B_2 - \{y\}) \cup \{x\} \in \mathcal{B}$, contradicting the maximality of $|B_1 \cap B_2|$. Therefore, $(E, \mathcal{I})$ is a matroid.

It remains to be shown that the family of maximal independent sets of $(E, \mathcal{I})$ is actually $\mathcal{B}$, which follows directly from the definition of $\mathcal{I}$ in (3.2). A set B is maximally independent in $(E, \mathcal{I})$ if and only if it is contained in a member of $\mathcal{B}$ and $B \cup \{x\}$ is not contained in any member of $\mathcal{B}$ for all $x \in E - B$. This is equivalent to $B \in \mathcal{B}$. $\square$

Axiom (B2) is called the *base exchange axiom* and is a generalization of Propositions 2.4, 2.10, and 2.14.

Given the definition of matroids with respect to the family of bases, we can see that there exist simple combinatorial structures that are matroids. For example, it is easy to see that all nonempty subsets of cardinality k of a finite set E satisfy the basis axiom (B2) of Theorem 3.2.

Definition 3.6 (Uniform Matroid) A matroid isomorphic to the matroid $U_{k,n}$ with ground set $|E| = n$ and independence family

$$\mathcal{I} = \{X \subseteq E : |X| \leq k\},$$

will be called **uniform matroid**.

3.3 Circuits

Definition 3.7 (Circuits) Given an independence system $(E, \mathcal{I})$ the minimal dependent sets will be called **circuits**. The family of circuits will be denoted by $\mathcal{C}$.

The collection of circuits for some $X \subseteq E$ will be denoted by $\mathcal{C}(X)$ and is defined as

$$\mathcal{C}(X) = \{Y \subseteq X : Y \notin \mathcal{I}, Y - \{y\} \in \mathcal{I} \text{ for all } y \in Y\}. \tag{3.6}$$

A singleton $e \in E(M)$ that is a circuit, will be called a **loop** of M.

Theorem 3.3 (Circuit Axioms) *A collection* $\mathscr{C} \subseteq 2^E$ *is the set of circuits of a matroid* $M(E, \mathscr{I})$ *if and only if the following are satisfied:*

(C1) $\emptyset \notin \mathscr{C}$.
(C2) *If* $C_1, C_2 \in \mathscr{C}$ *and* $C_1 \subseteq C_2$ *then* $C_1 = C_2$.
(C3) *If* $C_1, C_2 \in \mathscr{C}$, $C_1 \neq C_2$ *and* $e \in C_1 \cap C_2$ *then there exists* $C_3 \in \mathscr{C}$ *such that* $C_3 \subseteq (C_1 \cup C_2) - \{e\}$.

Proof Let $\mathscr{C}$ be the set of circuits of the matroid $M(E, \mathscr{I})$, where $\mathscr{I}$ satisfies axioms (I1)-(I3) of Definition 3.1. We will show that $\mathscr{C}$ satisfies (C1)-(C3). Axiom (C1) follows from axiom (I1), while (C2) is true since if $C_1 \neq C_2$ then C_2 is not minimally dependent. Consider now any $C_1, C_2 \in \mathscr{C}$, $C_1 \neq C_2$ and $e \in C_1 \cap C_2$, and assume by contradiction that $(C_1 \cup C_2) - \{e\}$ does not contain a member of $\mathscr{C}$. By Definition 3.7 we must have $(C_1 \cup C_2) - \{e\} \in \mathscr{I}$. Since by (C2) none of the C_1 and C_2 is a subset of the other, we have that there exists at least two elements $x_2 \in C_2 - C_1$ and $x_1 \in C_1 - C_2$. If we take the maximum independent subset X of $C_1 \cup C_2$, since X cannot contain either C_1 or C_2, we can assume without loss of generality that $x_1, x_2 \in X$. Therefore, we have

$$|X| \leq |(C_1 \cup C_2) - \{x_1, x_2\}| < |(C_1 \cup C_2) - \{e\}|.$$

Applying axiom (I3) to the independent sets X and $(C_1 \cup C_2) - \{e\}$ we get an independent subset of $C_1 \cup C_2$ which is larger than X, a contradiction.

Assuming now that $\mathscr{C} \subseteq 2^E$ satisfies (C1)-(C3), we will show that for

$$\mathscr{I} = \{X \subseteq E : \text{ there does not exist } Y \in \mathscr{C} \text{ such that } Y \subseteq X\}, \qquad (3.7)$$

the set system $(E, \mathscr{I})$ is a matroid. Axiom (I1) is established by (C1) and the fact that the empty set does not contain any member of $\mathscr{C}$. Axiom (I2) follows directly from the definition of $\mathscr{I}$ in (3.7). Therefore, $(E, \mathscr{I})$ is an independence system. We will now show that for any $X \subseteq E$ the bases of X have the same cardinality, thus by Lemma 3.1 $(E, \mathscr{I})$ is a matroid. Consider any $X \subseteq E$ and assume by contradiction that there exist $X_1, X_2 \in \mathscr{B}(X)$ such that $|X_1| > |X_2|$, while we choose those X_1 and X_2 such that $|X_1 \cap X_2|$ is maximum. Take any $x \in X_1 - X_2$ and consider the set $X_2 \cup \{x\}$. Since $X_2 \in \mathscr{B}(X)$ we have that $X_2 \cup \{x\} \notin \mathscr{I}$, which by (3.7) implies that it contains some $C_x \in \mathscr{C}$, where $x \in C_x$. If there exists some C distinct from C_x such that $C \in \mathscr{C}$ and $C \subseteq X_2 \cup \{x\}$, by (C3) we have that $(C \cup C_x) - \{x\}$ contains a member of $\mathscr{C}$, which is a contradiction since $(C \cup C_x) - \{x\} \subseteq X_2$. We can conclude that C_x is the unique circuit in $X_2 \cup \{x\}$. So if we take some $y \in C_x - X_1$ then $X_3 = (X_1 \cup \{x\}) - \{y\} \in \mathscr{I}$ since we eliminate the unique circuit in $X_2 \cup \{x\}$. But now we have $|X_3 \cap X_1| > |X_2 \cap X_1|$, which is a contradiction. Therefore, $(E, \mathscr{I})$ is a matroid.

It remains to be shown that the family of minimal dependent sets of $(E, \mathscr{I})$ is actually $\mathscr{C}$. This follows from the definition of $\mathscr{I}$ in (3.7), since C is a circuit of $(E, \mathscr{I})$ if and only if C contains a member of $\mathscr{C}$ while $C - \{x\}$ does not, for any $x \in C$, which is equivalent to $C \in \mathscr{C}$. $\qquad\square$

Axiom (C3) is called the *circuit elimination axiom* and is a generalization of Propositions 2.5, 2.11, and 2.15.

In the second part of the proof in Theorem 3.3, we show that for any independent set X in a matroid $M(E, \mathscr{I})$, if $X \cup \{e\} \notin \mathscr{I}$ for some $e \in E$, then $X \cup \{e\}$ contains a unique circuit, which contains e. The following proposition is an immediate result of this.

Proposition 3.1 *If $B \in \mathscr{B}(E)$ for a matroid $M(E, \mathscr{I})$ and $x \in E - B$, then there exists a unique circuit $C(x, B)$ contained in $B \cup \{x\}$ and it contains x.*

Any circuit in a matroid of the type defined in Proposition 3.1 will be called a **fundamental circuit** of a base. Next we provide a characterization for the elements in a fundamental circuit of a base.

Proposition 3.2 *If $B \in \mathscr{B}(E)$ for a matroid $M(E, \mathscr{I})$, then for any $x \in E - B$ the set $(B - \{y\}) \cup \{x\}$ is a base of M if and only if $y \in C(x, B)$.*

Proof Assume by contradiction that $(B - \{y\}) \cup \{x\}$ is a base of M for some $y \in B - C(x, B)$. Then $C(x, B)$ is contained in $(B - \{y\}) \cup \{x\}$, a contradiction. Consider now any $y \in C(x, B)$ and assume that $(B - \{y\}) \cup \{x\} \notin \mathscr{I}$. Then $B \cup \{x\}$ contains a circuit other than $C(x, B)$, which contradicts Proposition 3.1. □

We also state the following lemma that is used in the proof of the main result in Chap. 5.

Lemma 3.2 *If M_1 and M_2 are two matroids on the same ground set, such that any circuit of M_1 contains a circuit of M_2 and vice versa, then $M_1 = M_2$.*

Proof If $X \in \mathscr{C}(M_1)$ there exists a circuit $Y \in \mathscr{C}(M_2)$ such that $X \subseteq Y$ and for Y there exists a circuit $Z \in \mathscr{C}(M_1)$ such that $Z \subseteq Y$. We have

$$Z \subseteq Y \subseteq X,$$

and by axiom (C2) we must have $Z = X$, which implies that $X = Y$. Similarly, we show that any circuit of M_2 is a circuit of M_1, thus, both matroids have the same family of circuits and they are equal. □

3.4 Rank

Definition 3.8 (**Rank**) Given an independence system $(E, \mathscr{I})$, the **rank function** $r : 2^E \to \mathbb{Z}_+$ is defined as

$$r(X) = \max\{|Y| : Y \subseteq X, Y \in \mathscr{I}\} \tag{3.8}$$

for any $X \subseteq E$.

Theorem 3.4 (Rank Axioms) *A function* $r : 2^E \to \mathbb{Z}$ *is the rank function of a matroid* $M(E, \mathscr{I})$ *if and only if the following are satisfied for all* $X, Y \in E$:

(R1) $0 \leq r(X) \leq |X|$.
(R2) *If* $Y \subseteq X$ *then* $r(Y) \leq r(X)$.
(R3) $r(X) + r(Y) \geq r(X \cup Y) + r(X \cap Y)$.

Proof Let $r : 2^E \to \mathbb{Z}$ be the rank function of a matroid $M(E, \mathscr{I})$, where $\mathscr{I}$ satisfies axioms (I1)-(I3) of Definition 3.1. By Definition 3.8 r trivially satisfies (R1) and (R2). For (R3), first note that by Theorem 3.1 for any $X \subseteq E$ if $Y \subseteq X$ such that $Y \in \mathscr{I}$, then Y can be extended to a basis of X. Consider now a basis $A \in \mathscr{B}(X \cap Y)$. Then $A \in \mathscr{I}(X)$, so by Theorem 3.1 there must exist some $B \subseteq X$, such that $A \cap B = \emptyset$ and $(A \cup B) \in \mathscr{B}(X)$. Similarly, since $(A \cup B) \in \mathscr{I}(X \cup Y)$ there must exist some $C \subseteq X \cup Y$, $(A \cup B) \cap C = \emptyset$ and $(A \cup B) \cup C \in \mathscr{B}(X \cup Y)$. Therefore, $(A \cup C) \in \mathscr{I}$ and we can now state the following:

$$\begin{aligned} r(X) + r(Y) &\geq |A \cup B| + |A \cup C| \\ &= 2|A| + |B| + |C| \\ &= |A \cup B \cup C| + |A| \\ &= r(X \cup Y) + r(X \cap Y). \end{aligned}$$

Let now $r : 2^E \to \mathbb{Z}$ satisfy (R1)-(R3). We will show that for

$$\mathscr{I} = \{X \subseteq E : r(X) = |X|\}, \tag{3.9}$$

the set system $(E, \mathscr{I})$ is a matroid. By (R1) we have that $r(\emptyset) = 0$ which means that $\emptyset \in \mathscr{I}$. For (I2), take any $X \in \mathscr{I}$ and $Y \subseteq X$. Assume by contradiction that $Y \notin \mathscr{I}$. Then by (R1) and (3.9) we have that $r(Y) < |Y|$. By (R3) we have

$$\begin{aligned} r(X - Y) + r(Y) &\geq r((X - Y) \cup Y) + r((X - Y) \cap Y) \\ &= r(X). \end{aligned}$$

Since $X \in \mathscr{I}$ by assumption, and $r(X - Y) \leq |X - Y|$ by (R2), substituting the above we get

$$|X - Y| + |Y| > r(X - Y) + r(Y) \geq r(X) = |X|,$$

which is a contradiction. For (I3), assume by contradiction that there exist $X, Y \in \mathscr{I}$, $|X| > |Y|$, such that for all $x \in X - Y$, $Y \cup \{x\} \notin \mathscr{I}$. By (R1) and (3.9) we have that $r(Y \cup \{x\}) < |Y| + 1$ for all $x \in X - Y$. Moreover, by (R2) we have $r(Y \cup \{x\}) \geq r(Y)$, where $r(Y) = |Y|$ since $Y \in \mathscr{I}$ by assumption. Therefore, since r is an integer function we get

$$r(Y \cup \{x\}) = r(Y) = |Y|, \quad \text{for all } x \in X - Y. \tag{3.10}$$

Consider now two distinct $x_1, x_2 \in X - Y$. By (R3) we have

$$r(Y \cup \{x_1\}) + r(Y \cup \{x_2\}) \geq r((Y \cup \{x_1\}) \cup (Y \cup \{x_2\})) + r((Y \cup \{x_1\}) \cap (Y \cup \{x_2\}))$$
$$= r(Y \cup \{x_1, x_2\}) + r(Y).$$

So by (3.10) and (R2)

$$|Y| \geq r(Y \cup \{x_1, x_2\}) \geq r(Y),$$

which implies that $r(Y \cup \{x_1, x_2\}) = |Y|$. Thus, by repeated application of (R3) we will get $r(Y \cup (X - Y)) = |Y|$, which is a contradiction since $Y \cup (X - Y) = X$.

In order to show that the function $r : 2^E \to \mathbb{Z}$ which satisfies (R1)-(R3), is the rank function of $(E, \mathscr{I})$ as stated in Definition 3.8, we have to show that

$$r(X) = \max\{|Y| : Y \subseteq X, r(Y) = |Y|\},$$

for any $X \subseteq E$. Given any $X \subseteq E$, by (R1) and (R2) we have that Y is the maximum subset of X such that $r(Y) = |Y|$ if and only if for all $x \in X - Y, r(Y) \leq r(Y \cup \{x\}) < |Y| + 1$, which is equivalent to $r(Y \cup \{x\}) = |Y|$. By repeated application of (R3) as it is done above, we have $r(X) = r(Y)$. $\qquad\square$

Axiom (R3) is called the *submodularity* of r and it is a generalization of Propositions 2.7 and 2.12. There is also an equivalent set of rank axioms, as stated in Theorem 3.5 without proof.

Theorem 3.5 (Rank Axioms) *A function* $r : 2^E \to \mathbb{Z}$ *is the rank function of a matroid* $M(E, \mathscr{I})$ *if and only if the following are satisfied for all* $X \in E$ *and* $x, y \in E :$

(R1') $r(\emptyset) = 0$.
(R2') $r(X) \leq r(X \cup \{y\}) \leq r(X) + 1$.
(R3') *If* $r(X \cup \{x\}) = r(X \cup \{y\}) = r(X)$ *then* $r(X \cup \{x, y\}) = r(X)$.

Given a matroid $M(E, \mathscr{I})$ we can make the following observations regarding the rank function and the families of independent sets, bases, and circuits:

(i) $X \in \mathscr{I} \Leftrightarrow |X| = r(X)$.
(ii) $X \in \mathscr{B} \Leftrightarrow |X| = r(X) = r(E)$.
(iii) $X \in \mathscr{C} \Leftrightarrow X \neq \emptyset$ and for every $x \in X, r(X - \{x\}) = |X| - 1 = r(X)$.

For an independence system $(E, \mathscr{I})$ the rank of some $X \subseteq E$ is the cardinality of the largest base of X. By Lemma 3.1, if $(E, \mathscr{I})$ is a matroid then all bases of X have the same cardinality. If on the other hand $(E, \mathscr{I})$ is not a matroid, then X may contain bases which are less in size than $r(X)$.

Definition 3.9 (Low Rank) Given an independence system $(E, \mathscr{I})$, the **low rank function** $lr : 2^E \to \mathbb{Z}$ is defined as

$$lr(X) = \min\{|Y| : Y \subseteq X, Y \in \mathscr{I}, Y \cup \{x\} \notin \mathscr{I} \text{ for all } x \in Y - X\}$$

for any $X \subseteq E$.

So while rank is defined as the cardinality of the largest base of X, low rank is the cardinality of the smallest base. Note that $r(X) = |X| \Leftrightarrow lr(X) = |X|$, that is, rank and low rank are equivalent when X is an independent set. The importance of low rank will become evident in Sect. 3.7 when we will discuss the relationship between matroids and optimization. Given an independence system $(E, \mathscr{I})$ and $X \subseteq E$, the cardinality of X will lie in the integer interval $[0, |E|]$. Any X with size $r(E) < |X| \leq |E|$ is a dependent set. Any base $B \in \mathscr{B}$ will have size $lr(E) \leq |B| \leq r(E)$, while any X with size $0 \leq |X| < lr(X)$ can be either independent or dependent.

3.5 Closure

The closure of a set resembles the span of a set of vectors as defined in Sect. 2.2. Given an independence system $(E, \mathscr{I})$, a subset $X \subseteq E$ and $y \in E$, we will say that y **depends** on X and write $y \sim X$, when $r(X \cup \{y\}) = r(X)$. The closure of X is the set of those elements of E than depend on X.

Definition 3.10 (**Closure**) Given an independence system $(E, \mathscr{I})$ the **closure operator** is a set function $cl : 2^E \rightarrow 2^E$ defined as

$$cl(X) = \{y \in E : r(X \cup \{y\}) = r(X)\}. \tag{3.11}$$

for any $X \subseteq E$.

For some $X \subseteq E$, it follows from the definition that $X \subseteq cl(X)$, since $r(X \cup \{x\}) = r(X)$ for all $x \in X$. Moreover, it follows from the definition that

$$cl(X) = E \Leftrightarrow r(X) = r(E).$$

Example 3.2 If $M(E, \mathscr{I}) = M[A]$ for some matrix $A \in \mathbb{F}^{m \times n}$, then for $X \subseteq E$ we have
$$cl(X) = \{\mathbf{x} \in E : \mathbf{x} = \sum_i a_i \mathbf{x}_i, \mathbf{x}_i \in X, \text{ for some } a_i \in \mathbb{F}\},$$

or equivalently $cl(X) = \langle X \rangle \cap E$.

Consider now that we have the graphic matroid $M(G)$ of the graph G given in Fig. 2.1. In this case for $X = \{e_1, e_3\}$ we have $cl(X) = X \cup \{e_4, e_2\}$, while for $X = \{e_5, e_6\}$ we have $cl(X) = X \cup \{e_4, e_7, e_2\}$. Actually, since $r(e_2) = 0$ the loop e_2 is included in the closure of every subset of E.

For the transversal matroid of the set system $(E, \mathscr{F})$ in Example 2.1, for $X = \{1, 3, 5\}$ we have $cl(X) = X \cup \{4\}$ while for $X = \{1, 2\}$ we have $cl(X) = X$. $\square$

Theorem 3.6 (Closure Axioms) *A function* $cl : 2^E \to 2^E$ *is the closure operator of a matroid* $M(E, \mathscr{I})$ *if and only if the following are satisfied for all* $X, Y \subseteq E$ *and* $x, y \in E$:

(CL1) *If* $X \subseteq E$ *then* $X \subseteq cl(X)$.
(CL2) *If* $X \subseteq Y \subseteq E$ *then* $cl(X) \subseteq cl(Y)$.
(CL3) *If* $X \subseteq E$ *then* $cl(cl(X)) = cl(X)$.
(CL4) *If* $X \subseteq E$, $x \in E$, $y \in cl(X \cup \{x\}) - cl(X)$ *then* $x \in cl(X \cup \{y\})$.

Proof Let $cl : 2^E \to 2^E$ be the closure operator of a matroid $M(E, \mathscr{I})$, where $\mathscr{I}$ satisfies axioms (I1)-(I3) of Definition 3.1. Axiom (CL1) follows from Definition 3.10. For (CL2) let $X \subseteq Y \subseteq E$ and take any $x \in cl(X)$. If $x \in X$ then $x \in Y$ and $x \in cl(Y)$. If $x \notin X$, then since any base of X can be extended to a base of Y by axiom (I3), we can find a base of Y that does not contain x. This implies that $r(Y \cup \{x\}) = r(Y)$, or $x \in cl(Y)$. Thus, we have shown that $cl(X) \subseteq cl(Y)$. For (CL3) since by (CL1) we have that $cl(X) \subseteq cl(cl(X))$, enough to show that $cl(cl(X)) \subseteq cl(X)$. Take any $x \in cl(cl(X))$, where by the definition of closure we know that $r(cl(X) \cup \{x\}) = r(cl(X))$. Since $cl(X) = X \cup cl(X) - X$, we have

$$
\begin{aligned}
r(X) &= r(X \cup cl(X) - X) \\
&= r(cl(X)) \\
&= r(cl(X) \cup \{x\}).
\end{aligned}
\tag{3.12}
$$

By axiom (R2) and the fact that $X \cup \{x\} \subseteq cl(X \cup \{x\})$ we have

$$
r(cl(X) \cup \{x\}) \geq r(X \cup \{x\}) \geq r(X).
$$

By (3.12) we get $r(X \cup \{x\}) = r(X)$ and $x \in cl(X)$. For (CL4) let $y \in cl(X \cup \{x\}) - cl(X)$ for some $X \subseteq E$ and $x \in E$. This implies that $r(X \cup \{x, y\}) = r(X \cup \{x\})$ and $r(X \cup \{y\}) \neq r(X)$. Now for any $X \subseteq E$ and $x \in E$ we have

$$
r(X) \leq r(X \cup \{x\}) \leq r(X) + 1,
$$

which by the integrality of r means that either $r(X) = r(X \cup \{x\})$ or $r(X \cup \{x\}) = r(X) + 1$. We can write then $r(X \cup \{y\}) = r(X) + 1$. Therefore, we have

$$
r(X) + 1 = r(X \cup \{y\}) \leq r(X \cup \{y\} \cup \{x\}) = r(X \cup \{x\}) \leq r(X) + 1,
$$

and the inequality collapses resulting in $r(X \cup \{y\} \cup \{x\}) = r(X \cup \{y\}) \Leftrightarrow x \in cl(X \cup \{y\})$.

Let $cl : 2^E \to 2^E$ satisfy (CL1)-(CL4). We will show that for

$$
\mathscr{I} = \{X \subseteq E : x \notin cl(X - \{x\}) \text{ for all } x \in X\},
\tag{3.13}
$$

the set system $(E, \mathscr{I})$ is a matroid. Trivially, we have $\emptyset \in \mathscr{I}$. For (I2), consider
some $X \in \mathscr{I}$ and $Y \subseteq X$. Since $Y - \{x\} \subseteq X - \{x\}$, by (CL2) we have $cl(Y - \{x\}) \subseteq$
$cl(X - \{x\})$. Thus if $x \notin cl(X - \{x\})$ then $x \notin cl(Y - \{x\})$ for all $x \in Y$, which
means that $Y \in \mathscr{I}$.

To prove that (I3) holds, assume by contradiction that there exist $X, Y \in \mathscr{I}$ such
that $|X| > |Y|$, while for all $x \in X - Y$ we have that $Y \cup \{x\} \notin \mathscr{I}$. From all such X, Y
take those that $|X \cap Y|$ is maximum. Consider the set $X - \{x\}$ for some $x \in X - Y$.
Since we have shown already that (I2) is true, $X - \{x\} \in \mathscr{I}$. If $Y \subseteq cl(X - \{x\})$
then by (CL2)

$$cl(Y) \subseteq cl(cl(X - \{x\})) = cl(X - \{x\}),$$

thus, since $x \notin cl(X - \{x\})$ because $X \in \mathscr{I}$, we have $x \notin cl(Y)$. This implies
that $Y \cup \{x\} \in \mathscr{I}$ by the definition of $\mathscr{I}$ in (3.13), which is a contradiction. So let
$Y \nsubseteq cl(X - \{x\})$. There must exist some $y \in Y - X$ such that $y \notin cl(X - \{x\})$.
Since $X - \{x\} \in \mathscr{I}$ by (I2), we have that $(X - \{x\}) \cup \{y\} \in \mathscr{I}$ by the definition of
$\mathscr{I}$ in (3.13). But

$$|((X - \{x\}) \cup \{y\}) \cap Y| > |X \cap Y|,$$

which in turn implies that for Y and $(X - \{x\}) \cup \{y\}$ axiom (I3) is true, since we
assumed that $|X \cap Y|$ is maximum. Thus, there exists $z \in ((X - \{x\}) \cup \{y\}) - Y$
such that $Y \cup \{z\} \in \mathscr{I}$. But $z \in X - Y$ in this case, so (I3) is true for X, Y also, a
contradiction.

In order to show that $cl : 2^E \rightarrow 2^E$ which satisfies (CL1)–(CL4), is the closure
operator of $(E, \mathscr{I})$ as stated in Definition 3.10, we have to show that for any $X \subseteq E$
and $y \in E$

$$r(X \cup \{y\}) = r(X) \Leftrightarrow y \in cl(X).$$

Consider any $X \subseteq E$. For any $y \in X$, $y \in cl(X)$ by (CL1) and $r(X \cup \{y\}) = r(X)$.
For some $y \in E - X$, assume that $r(X \cup \{y\}) = r(X)$ and $y \notin cl(X)$. Then $y \notin cl(B)$
for some base $B \in \mathscr{B}(X)$ by (CL2) since $B \subseteq X$. This implies that $B \cup \{y\} \in \mathscr{I}$,
that is

$$z \notin cl((B \cup \{y\}) - \{z\}), \quad \text{for all } z \in B \cup \{y\},$$

for otherwise there must exist $z \in cl(B \cup \{y\}) - cl(B)$, and by (CL4) we will
have $y \in cl(B \cup \{y\}) = cl(B)$. Therefore, $r(X \cup \{y\}) = r(B \cup \{y\}) > r(X)$,
a contradiction. Similarly, it can be shown that if $y \in cl(X)$ then $r(X \cup \{y\})$
$= r(X)$.   □

Axiom (CL4) is also known as the *MacLane – Steinitz exchange* property.

Given an independence system $(E, \mathscr{I})$, if $X = cl(X)$ then X is called a **flat** or a
closed set. If X is a flat and $r(X) = r(E) - 1$ then we will call X a **hyperplane**,
while if $cl(X) = E$ then X is called a **spanning set** of the independence system.

Due to the equivalency of the axiom systems of the previous sections, whenever
we are referring to a matroid we will interchangeably use the definitions based on the

independent sets $M(E, \mathscr{I})$, family of bases $M(E, \mathscr{B})$, family of circuits $M(E, \mathscr{C})$, rank function $M(E, r)$, or closure operator $M(E, cl)$.

3.6 Dependent Sets, Spanning Sets, and Hyperplanes

The procedure that we employed in Sect. 3.2 for establishing equivalence between the two definitions of matroids, with respect to the axiom systems (I1)–(I3) and (B1), (B2) that correspond to to the families of independent sets and bases, is the following. An *interpretation* is provided in (3.1) to construct $\mathscr{B}$ from $\mathscr{I}$ and in (3.2) to construct $\mathscr{I}$ from $\mathscr{B}$. Based on these interpretations, in Theorem 3.2 we proved that one axiom system implies the other. Similarly for the family of circuits, the rank function and the closure operator, in Sects. 3.3, 3.4 and 3.5 respectively. In this section we will provide similar interpretations between the family of independent sets and families of dependent sets, spanning sets, and hyperplanes. The set of axioms that define a matroid with respect to each family will also be provided.

The family $\mathscr{D}$ of dependent sets is complementary to the family of independent sets, so we have

$$\mathscr{I} = \{X \subseteq E : X \notin \mathscr{D}\}.$$

Theorem 3.7 (Dependent Sets Axioms) *A collection $\mathscr{D} \subseteq 2^E$ is the set of dependent sets of a matroid $M(E, \mathscr{I})$ if and only if the following are satisfied:*

(D1) $\emptyset \notin \mathscr{D}$.
(D2) *If $X \in \mathscr{D}$ and $X \subseteq Y$ then $Y \in \mathscr{D}$.*
(D3) *If $X, Y \in \mathscr{D}$ and $X \cap Y \notin \mathscr{D}$, then for every $x \in E$, $(X \cup Y) - \{x\} \in \mathscr{D}$.*

Comparing the spanning set axioms in Theorem 3.8 with the independent sets axioms in Definition 3.1, we notice a *duality* relationship, which will become evident in Sect. 4.2.

Theorem 3.8 (Spanning Sets Axioms) *A collection $\mathscr{S} \subseteq 2^E$ is the set of spanning sets of a matroid $M(E, \mathscr{I})$ if and only if the following are satisfied:*

(S1) $\mathscr{S} \neq \emptyset$.
(S2) *If $X \in \mathscr{S}$ and $X \subseteq Y$ then $Y \in \mathscr{S}$.*
(S3) *If $X, Y \in \mathscr{S}$ and $|X| > |Y|$, then there exists $x \in X - Y$ such that $X - \{x\} \in \mathscr{S}$.*

Given the family of hyperplanes, we can define the family of independent sets as follows:

$$\mathscr{I} = \{X \subseteq E : \text{for all } x \in X, X - H = \{x\} \text{ for some } H \in \mathscr{H}\},$$

while the definition of matroids based on hyperplanes is the following.

Theorem 3.9 (Hyperplane Axioms) *A collection $\mathscr{H} \subseteq 2^E$ is the set of hyperplanes of a matroid $M(E, \mathscr{I})$ if and only if the following are satisfied:*

(H1) $E \notin \mathcal{H}$.
(H2) If $H_1, H_2 \in \mathcal{H}$ and $H_1 \subseteq H_2$ then $H_1 = H_2$.
(H3) If $H_1, H_2 \in \mathcal{H}$ and H_1 distinct from H_2, then for every $x \in E$ there exists $H_3 \in \mathcal{H}$ such that $(H_1 \cap H_2) \cup \{x\} \subseteq H_3$.

3.7 Greedy Algorithm

Although in this section we will provide another equivalent definition of matroids, it is unique in the sense that it is not based on a set of axioms, but rather on the output of an algorithm to an optimization problem. This *algorithmic* definition of matroids that will be given in Theorem 3.11, is a rare instance of a mathematical entity being characterized by an algorithm. Moreover, the nature of the definition establishes the fundamental role that matroids have in optimization.

Consider an independence system $(E, \mathcal{I})$ together with a weight function $w : E \to \mathbb{R}$ which is linear, that is, for any $X \subseteq E$ we have

$$w(X) = \sum_{e \in X} w(e).$$

We can state the following general discrete optimization problem on independence systems

$$\max(\min)\quad w(X) \tag{3.14}$$
$$\text{s.t.}\quad X \in \mathcal{B},$$

where $w(\emptyset) = 0$ and $\mathcal{B}$ is the family of bases of E. Any $X \in \mathcal{B}$ will be called a **feasible** solution, while if it also solves (3.14) it will be called an **optimum** solution. Most known combinatorial optimization problems can be stated as optimization problems on independence systems, where the objective is to find a minimum or a maximum weight basis. Consider for instance the following combinatorial optimization problems on graphs, where in each problem it is easy to show that $\mathcal{I}$ satisfies axioms (I1) and (I2) of Definition 3.1. In each case we are given a connected undirected graph $G(V, E)$.

1. *Minimum Spanning Tree.* For some weight function $w : E(G) \to \mathbb{R}$ find a spanning tree T of G such that $w(E(T))$ is minimum. Set $E = E(G)$ and $\mathcal{I} = \{X \subseteq E : G[X] \text{ is a forest}\}$.
2. *Maximum Weight Matching.* For some weight function $w : E(G) \to \mathbb{R}$ find a matching $M \subseteq E(G)$ such that $w(M)$ is maximum. Set $E = E(G)$ and $\mathcal{I} = \{X \subseteq E : X \text{ is a matching}\}$.
3. *Traveling Salesman Problem.* For some weight function $w : E(G) \to \mathbb{R}_+$ find a spanning cycle C of G such that $w(E(C))$ is minimum. Set $E = E(G)$ and $\mathcal{I} = \{X \subseteq E : X \subseteq E(C), C \text{ is a spanning cycle of G}\}$.

4. *Maximum Clique Problem.* For a constant weight function $w(v) = 1$ for all $v \in V(G)$, find $T \subseteq V(G)$ such that $G[T]$ is a complete graph and $w(T)$ maximum. Set $E = V(G)$ and $\mathscr{I} = \{X \subseteq E : G[X] = K_{|X|}\}$.

For the minimum spanning tree problem, we know that the corresponding independence system $(E, \mathscr{I})$ is a graphic matroid since $\mathscr{I}$ satisfies axiom (I3) as we have shown in Proposition 2.3. However, for the other problems it is easy to construct a counterexample upon which axiom (I3) fails, as we have done in Example 3.1 for the maximum weight matching problem. For those independent systems that are not matroids, we have also defined the low rank in Definition 3.9, which is the cardinality of the smallest base in a set. Consider now the ratio

$$q(E, \mathscr{I}) = \min_{X \subseteq E} \frac{lr(X)}{r(X)}, \qquad (3.15)$$

which will be called the **rank quotient** of $(E, \mathscr{I})$. The rank quotient is directly proportional to difference between the smallest and largest base of an independent system, so in that sense, it could be used as a measure of how *close* the independent system is from having a matroidal structure.

Perhaps the most naive algorithm for solving (3.14) is the so-called greedy algorithm, which is described in Algorithm 3.1 for the maximization problem. The

---

Algorithm 3.1 GREEDY

Input : independence system $(E, \mathscr{I})$, function $w : E \to \mathbb{R}$
Output: set $X \in \mathscr{B}$

1. Sort E such that $w(e_1) \geq w(e_2) \geq \ldots \geq w(e_{|E|})$
2. $X := \emptyset$
3. **for** $i = 1, \ldots, |E| \to$
4.   **if** $X \cup \{e_i\} \in \mathscr{I} \to$
5. $X := X \cup \{e_i\}$
6.   **end if**
7. **end for**
8. **return** X

---

correctness of the GREEDY, that is, the fact that it always produces a base of $(E, \mathscr{I})$, is an immediate consequence of axiom (I2) and line 4 of the algorithm. We can see that the GREEDY algorithm starts from an empty solution, which by axiom (I1) is independent, and inserts the best elements into a partial solution until it becomes dependent. The procedure is myopic, in the sense that it never backtracks. Apparently the computationally intensive task in Algorithm 3.1 is in line 4, where we have to check whether some set X is independent or not. The corresponding GREEDY algorithm for the minimization problem in (3.14), is a slight modification of Algorithm 3.1 where the order upon which the ground set E is sorted in line 1 is reversed.

The following result demonstrates that the approximation performance of the GREEDY algorithm is indeed bounded by the rank quotient of the corresponding independence system.

Theorem 3.10 (Hausmann et al. (1980)) *Let $(E, \mathscr{I})$ be an independence system and $w : E \to \mathbb{R}$ any weight function. If for the maximization problem in (3.14) G is the solution found by the* GREEDY *algorithm and O is the optimum solution then*

$$q(E, \mathscr{I}) \leq \frac{w(G)}{w(O)}. \tag{3.16}$$

Moreover, the bound is sharp.

Proof Let $E_i = \{e_k : k = 1, \ldots, i\}$ for all $i = 1, 2, \ldots, n$. Moreover, let $O_i = O \cap E_i$ and $G_i = G \cap E_i$, while it is understood that $O_n = O$ and $G_n = G$. Observe that

$$|G_i| - |G_{i-1}| = \begin{cases} 1 & \text{if } e_i \in G, \\ 0 & \text{otherwise,} \end{cases} \tag{3.17}$$

and similarly for O_i. Since $O_i \subseteq E_i$ and $O_i \in \mathscr{I}$ we have by the definition of the rank function that

$$|O_i| = r(O_i) \leq r(E_i). \tag{3.18}$$

By line 4 of Algorithm 3.1 and axiom (I2), G_i is a basis for E_i for every $i = 1, \ldots n$, therefore,

$$|G_i| \geq lr(E_i). \tag{3.19}$$

Combining (3.17), (3.18) and (3.19), the fact that $w(e_i) - w(e_{i+1}) \geq 0$ for all $i = 1, \ldots n$, and setting $w(e_{n+1}) = 0$, we can relate the total weight of G and O as follows:

$$
\begin{aligned}
w(G) &= \sum_{i=1}^{n} (|G_i| - |G_{i-1}|) w(e_i) \\
&= \sum_{i=1}^{n} |G_i| w(e_i) - \sum_{i=1}^{n} |G_{i-1}| w(e_i) \\
&= \sum_{i=1}^{n} |G_i| w(e_i) - \sum_{i=1}^{n} |G_i| w(e_{i+1}) \\
&= \sum_{i=1}^{n} |G_i| (w(e_i) - w(e_{i+1})) \\
&\geq \sum_{i=1}^{n} lr(E_i)(w(e_i) - w(e_{i+1}))
\end{aligned}
$$

$$\geq q(E, \mathscr{I}) \sum_{i=1}^{n} r(E_i)(w(e_i) - w(e_{i+1}))$$

$$\geq q(E, \mathscr{I}) \sum_{i=1}^{n} |O_i|(w(e_i) - w(e_{i+1}))$$

$$= q(E, \mathscr{I}) \sum_{i=1}^{n} (|O_i| - |O_{i-1}|)w(e_i)$$

$$= q(E, \mathscr{I})w(O).$$

In order to show that the bound is sharp, we have to construct a weight function $w : E \to \mathbb{R}$ for any independence system, such that (3.16) holds with equality. For any $(E, \mathscr{I})$ take the set $X \subseteq E$ responsible for the value of the rank quotient of the independent system, that is, there exist $B_1, B_2 \in \mathscr{B}(X)$ such that

$$\frac{|B_1|}{|B_2|} = q(E, \mathscr{I}).$$

It is enough to make B_1 the solution found by the GREEDY algorithm and B_2 the optimum for the maximization problem in (3.14). This can be easily done if we consider the weight function

$$w(e) = \begin{cases} 1 & \text{if } e \in X, \\ 0 & \text{otherwise,} \end{cases}$$

and take any ordering of E where the first elements are those of B_1. ☐

It is easy to see that the performance of the GREEDY algorithm for the minimization problem in (3.14) is not bounded. However, in (Hausmann et al. 1980) a *dual* variant of the Algorithm 3.1 is presented, which has similar performance bounds as those given in Theorem 3.10 for the minimization problem.

Example 3.3 Assume that we are interested in finding a *maximum* weight matching in the graph $K_{2,2}$ illustrated in Fig. 3.2. The corresponding independent system will be described by $E = \{e_1, e_2, e_3, e_4\}$ and the family of bases as

$$\mathscr{B} = \{\{e_1, e_4\}, \{e_2, e_3\}\}.$$

If we have a weight function $w : E \to \mathbb{R}$ such that $w(e_3) = 0$ and

$$w(e_2) \geq w(e_1) \geq w(e_4) \geq 0,$$

then the solution that the GREEDY algorithm will give is $\{e_2, e_3\}$ with total weight

$$w(e_2) \geq \frac{1}{2}(w(e_1) + w(e_4)).$$

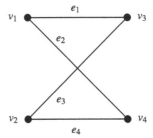

Fig. 3.2 Assignment problem

This is because for $X = \{e_1, e_2, e_4\}$ we have

$$\frac{lr(X)}{r(X)} = \frac{1}{2} = q(E, \mathscr{I}),$$

while it can be easily verified that this is the case for any $K_{n,n}$ since the size of any circuit in the corresponding independent system will be two. Any set of two edges incident to the same vertex is a circuit. On the other hand, if we were interested in finding a minimum weight matching in the graph $K_{2,2}$ of Fig. 3.2, we can see that the ratio between the solution produced by the GREEDY algorithm and the optimal is unbounded. Set for instance $w(e_3) = 0$, $w(e_1) = w(e_4) = 1$ and $w(e_2) = \infty$.   □

In view of Lemma 3.1, Theorem 3.10 yields the following algorithmic characterization of matroids.

Theorem 3.11 (Edmonds (1971)) *An independence system $(E, \mathscr{I})$ is a matroid if and only if the GREEDY algorithm is optimal for the maximization problem in (3.14).*

Proof If $(E, \mathscr{I})$ is a matroid, then by Lemma 3.1 we have $q(E, \mathscr{I}) = 1$, and by Theorem 3.10 the GREEDY is optimal.

Let $w(e) = 1$ for all $e \in E$ and take any $X \subseteq E$. It can be easily verified that for $\mathscr{I}_X = \{Y \subseteq X : Y \in \mathscr{I}\}$ the set system $(X, \mathscr{I}_X)$ is an independence system. Consider now the GREEDY algorithm for the maximization problem in (3.14) for the independence system $(X, \mathscr{I}_X)$. Since the weight function is constant, the ground set X can be ordered arbitrarily, which implies that the GREEDY can provide as a solution any base $B \in \mathscr{B}(X)$. Since the GREEDY is optimal, this implies that all bases of X must have cardinality $r(X)$, and by Lemma 3.1 the independence system $(E, \mathscr{I})$ is a matroid.   □

The rank quotient $q(E, \mathscr{I})$ of an independence system can be considered as a measure of the degree of matroid structure that the independence system exhibits, and as it is stated in Theorem 3.10 it provides a bound on the approximate solution found by the GREEDY on the associated maximization problem. However, its definition in

(3.15) does not seem to offer a direct way to efficiently compute its value for a given independence system. In what follows, we will provide a bound on the rank quotient that can be easily obtained, once the independence system has been defined in terms of matroids. The following proposition is easy to show.

Proposition 3.3 *Given matroids $M_1, \ldots, M_n$ on a common ground set E, the set system $(E, \mathscr{I})$ where*

$$\mathscr{I} = \{X \subseteq E : X \in \mathscr{I}(M_i), i = 1, \ldots, n\}$$

is an independence system.

We will call the independence system $(E, \mathscr{I})$ the **intersection** of the matroids $M_1, \ldots, M_n$ and write $M_1 \cap \cdots \cap M_n$. It turns out that any independence system is the intersection of a finite number of matroids.

Proposition 3.4 *If $(E, \mathscr{I})$ is an independence system, then there exist matroids $M_1, \ldots, M_n$ on E such that for any $X \subseteq E$*

$$X \in \mathscr{I} \Leftrightarrow X \in \mathscr{I}(M_1 \cap \cdots \cap M_n).$$

Proof Let $\mathscr{C} = \{C_1, \ldots, C_n\}$ be the family of circuits of $(E, \mathscr{I})$ and for each $C_i \in \mathscr{C}$ define the family

$$\mathscr{I}_{C_i} = \{X \subseteq E : C_i \not\subseteq X\}.$$

We will show that the set system $(E, \mathscr{I}_C)$ is a matroid for each $C \in \mathscr{C}$. The independence axioms (I1) and (I2) are clearly satisfied. For (I3) consider any $X, Y \in \mathscr{I}_C$ with $|X| > |Y|$, and assume that for all $x \in X - Y$, $C \subseteq Y \cup \{x\}$. We have that $X - Y \subseteq C$. If $|X - Y| \geq 2$ then $C \not\subseteq Y \cup \{x\}$ for all $x \in X - Y$, therefore, $X - Y = \{e\}$. Then $C \subseteq Y \cup \{e\} = X$, a contradiction.

Thus, $M_i(E, \mathscr{I}_{C_i})$ is a matroid for $i = 1, \ldots, n$, and we have

$$X \in \mathscr{I}(M_1 \cap \cdots \cap M_n) \Leftrightarrow X \text{ contains no circuit from } \mathscr{C} \Leftrightarrow X \in \mathscr{I}.$$

$\square$

The following theorem can be thought of as a generalization of Proposition 3.1 for independent systems.

Theorem 3.12 (Hausmann et al. (1980)) *Let $(E, \mathscr{I})$ be an independence system. If for every $X \in \mathscr{I}$ and $e \in E - X$ the set $X \cup \{e\}$ contains at most p circuits, then*

$$q(E, \mathscr{I}) \geq \frac{1}{p}.$$

Proof Consider any $X \subseteq E$ and $B_1, B_2 \in \mathscr{B}(X)$. We have to show that

$$\frac{|B_1|}{|B_2|} \geq \frac{1}{p}.$$

Let $B_1 - B_2 = \{b_1, \ldots, b_t\}$. The set $K_1 = B_2 \cup \{e_1\}$ contains at most p circuits by assumption and each circuit must contain $\{e_1\}$ and at least one element from $B_2 - B_1$ since $(B_1 \cap B_2) \cup \{e_1\} \in \mathscr{I}$. Denote by $X_1 \subseteq B_2 - B_1$ the set that contains one element from each circuit, where we have that $|X_1| \leq p$, since otherwise $B_2 \cup \{e_1\}$ can contain more than p circuits. Now let $K_i = (K_{i-1} - X_{i-1}) \cup \{e_i\}$ and X_i be computed in the same way, for $i = 1, \ldots, t$. We will have

$$|B_2 - B_1| = \left| \bigcup_{i=1}^{t} X_i \right| \leq pt = p|B_1 - B_2| \Rightarrow$$

$$|B_2| - |B_1 \cap B_2| = |B_2 - B_1| \leq p(|B_1| - |B_1 \cap B_2|) \Rightarrow$$

$$|B_2| \leq p|B_1| + |B_1 \cap B_2|(1 - p) \leq p|B_1| \Rightarrow$$

$$\frac{|B_1|}{|B_2|} \geq \frac{1}{p}.$$

$\square$

We have the following corollary from Theorem 3.12.

Corollary 3.1 *If the independence system $(E, \mathscr{I})$ is the intersection of p matroids then*

$$q(E, \mathscr{I}) \geq \frac{1}{p}.$$

Proof Let $(E, \mathscr{I})$ be the intersection of matroids $M_1, \ldots, M_p$ on E. By Proposition 3.1, if $X \in \mathscr{I}(M_1 \cap \cdots \cap M_p)$ and $e \in E$ then $X \cup \{e\}$ contains at most one circuit in each $M_i, i = 1, \ldots, n$. Therefore, X contains at most p circuits in $(E, \mathscr{I})$ and the result follows from Theorem 3.12. $\square$

Combined with Theorem 3.10, Corollary 3.1 provides an approximation ratio of the solution provided by GREEDY algorithm for the maximization problem in (3.14) and the optimum solution, with respect to the number of matroids used to describe the independence system.

Example 3.4 Consider the maximum weight bipartite matching problem, an instance of which was discussed in Example 3.3. We are given a bipartite graph $G(V_1, V_2, E)$ where $V(G) = V_1 \cup V_2$, and a nonnegative weight function $w : E \to \mathbb{R}$. Each vertex set defines a partition on the edges of the graph. Let the partition of E defined by the vertex set V_1 be the family $\mathscr{S} = (S_v : v \in V_1)$ where

$$S_v = \{e \in E(G) : e \text{ incident to } v\},$$

while let the partition defined by V_2 be the family $\mathscr{T} = (T_v : v \in V_2)$ where

$$T_v = \{e \in E(G) : e \text{ incident to } v\}.$$

It follows that the independence system that describes the maximum weight matching in $G(V_1, V_2, E)$, is the intersection of the transversal matroids $(E, \mathscr{S})$ and $(E, \mathscr{T})$. Therefore, as illustrated in Example 3.3, the GREEDY algorithm will have an approximation ratio of $\frac{1}{2}$. This is readily generalizable to k-partite matching, also known as multidimensional assignment problems. □

3.8 Notes

The main references for Sects. 3.1–3.5 are the books of Oxley (1992) and Welsh (1976). The proof of Theorem 3.6 is by Oxley (1992). Theorems 3.10 and 3.12 in Sect. 3.7 are from Hausmann et al. (1980).

The pairs of the equivalent axiom systems from Sects. 3.1–3.6 are called **cryptomorphic**, while the related interpretations **cryptomorphisms**. There are at least 13 known cryptomorphic axiom systems for matroids, which provide equivalent axiomatic definitions with respect to the families of non-spanning sets, open sets, and flats among others. These are listed by Brylawski (1986), where the author also provides interpretations of the corresponding families to graphs, vector spaces, and transversals. Moreover, in the same volume there is also a chapter by Nicoletti and White (1986), where equivalence is proved for nine of these, as we did here in Theorems 3.2–3.6.

The greedy algorithm is one of the oldest and simplest algorithms in optimization. The term *greedy* was coined by Edmonds (1971) for discrete optimization problems, but the algorithm is also encountered with other names such as *steepest descent* in continuous optimization. There have been several attempts in the past with varying degrees of success, to fully characterize the family of problems for which the greedy algorithm provides the optimal solution. Theorem 3.11 by Edmonds (1971) was also established independently by Rado (1957) and Gale (1968). Korte and Lovász (1981) considered introducing order into the independence systems and defined greedoids which generalize matroids. They have provided necessary and sufficient conditions upon which the greedy algorithm produces the optimal solution on greedoids. Faigle (1979) characterized those independence systems on partially ordered sets, where the greedy algorithm is optimal. Probably the most complete characterization of the problem structure where the greedy is optimal for linear objective functions is that of matroid embeddings, introduced by Helman et al. (1993). Theorem 3.10 first appeared in Jenkyns (1976), and the proof given here is by Korte and Hausmann (1978). Furthermore, an extension of the greedy algorithm to examine more than one element while constructing a solution is given by Korte and Hausmann (1978), while an analysis of its worst case performance for independence systems which are not necessarily matroids is given by Hausmann et al. (1980).

Chapter 4
Representability, Duality, Minors, and Connectivity

The material in this chapter constitutes a brief look at what can be considered as the fundamental core of matroid theory. We will present equivalent matrix representations of graphic matroids, the concept of duality, minors, and connectivity, and show how these apply to the two main classes of matroids presented in this book, namely, graphic and representable matroids. The focus will be on results needed for the exposition that follows in Chap. 5.

4.1 Representability

We have seen that graphs, matrices, and transversals can be thought of as *representations* of the same abstract object called matroid, in different settings. Any graph, or matrix, has a well-defined matroid associated with it, and namely, the cycle and vector matroid respectively. Assume now that we are given a matroid not by a graph, or a matrix, but by a ground set E and an **independence oracle**, that is, a procedure that provides an answer on whether some set $X \subseteq E$ is independent or not. We could define similar oracles for bases, circuits, rank, etc. Given a matroid by an independence oracle it is not clear what, if any, representation there exists of this matroid as a graph or a matrix. In Chap. 5 we will answer this question for graphic matroids, by establishing a decomposition theorem that could be used to decide in polynomial time whether a given binary matroid has a representation as a graph. In this section, we will demonstrate that graphic matroids are representable on any field. For matrices, however, representability is a much harder question to resolve. Establishing conditions upon which a matroid is representable by a matrix in some field $\mathbb{F}$, is a deep and rich area of research in matroid theory.

Let us associate a matrix with any given graph. The **incidence matrix** of a graph $G(V, E)$, is a matrix $A_G = (a_{ij}) \in GF(2)$ defined by

$$a_{ij} = \begin{cases} 1 & \text{if non-loop edge } e_j \text{ is incident to vertex } v_i, \\ 0 & \text{otherwise.} \end{cases}$$

L. S. Pitsoulis, *Topics in Matroid Theory*,
SpringerBriefs in Optimization, DOI: 10.1007/978-1-4614-8957-3_4,
© Leonidas S. Pitsoulis 2014

For example, the graph in Fig. 2.1 has the following incidence matrix

$$
A_G = \begin{array}{c} \\ v_1 \\ v_2 \\ v_3 \\ v_4 \end{array}
\begin{array}{c} e_1\ e_2\ e_3\ e_4\ e_5\ e_6\ e_7 \\
\left[\begin{array}{ccccccc}
1 & 0 & 0 & 1 & 1 & 0 & 0 \\
1 & 0 & 1 & 0 & 0 & 0 & 0 \\
0 & 0 & 1 & 1 & 0 & 1 & 1 \\
0 & 0 & 0 & 0 & 1 & 1 & 1
\end{array}\right]. \end{array}
\tag{4.1}
$$

Observe that for loop-less graphs, there is a one-to-one correspondence between graphs and matrices in $GF(2)$ with exactly two nonzero elements in each column. A zero column in such matrix corresponds to a loop in the graph, but the end-vertex of the loop is not uniquely determined. However, as we shall see in Sect. 4.4 this is of no importance in our context, since a loop may be placed in any vertex of a graph without affecting the corresponding cycle matroid. The next theorem shows that the incidence matrix of a graph G is a representation of $M(G)$ in $GF(2)$.

Theorem 4.1 *Graphic matroids are GF(2)-representable.*

Proof Let A_G be the incidence matrix of a graph $G(V, E)$. We shall prove that $M(G) = M[A_G]$ by showing that X is a linearly dependent set of columns in A_G if and only if $G[X]$ contains a cycle in G. Note that in $GF(2)$ a set of vectors X is linearly dependent if there exists some $\{\mathbf{x}_1, \dots, \mathbf{x}_k\} \subseteq X$ for $k \geq 1$ such that $\mathbf{x}_1 + \cdots + \mathbf{x}_k = \mathbf{0}$.

Assume that $G[X]$ contains a cycle $G[C]$ for some $C = \{e_1, \dots, e_n\} \subseteq X$. The corresponding columns $\{\mathbf{a}_1, \dots, \mathbf{a}_n\}$ in A_G form the submatrix A_C, which is the incidence matrix of the cycle C, and we must have

$$
\sum_{i=1}^{n} \mathbf{a}_i = \mathbf{0},
\tag{4.2}
$$

since each row of A_C has exactly two ones. Assume now that X is a linearly dependent set of columns in A_G. Thus, there exists some $C \subseteq X$ such that (4.2) is true. This in turn implies that the submatrix of A_G formed by the columns in C, has no row with odd number of nonzero elements, which means that the subgraph $G[C]$ has no vertex of degree one, therefore by Proposition 2.2 it is not a forest, and it must contain a cycle. □

Consider now that we have a directed graph $\overrightarrow{G}(V, E)$. The incidence matrix of $\overrightarrow{G}$ is a matrix $A_{\overrightarrow{G}} = (a_{ij}) \in \mathbb{R}$ defined by

$$
a_{ij} = \begin{cases}
+1 & \text{if vertex } i \text{ is the head of the non-loop arc } j, \\
-1 & \text{if vertex } i \text{ is the tail of the non-loop arc } j, \\
0 & \text{otherwise.}
\end{cases}
$$

Since each column in an $m \times n$ incidence matrix of a directed graph contains zero or two nonzero entries with different sign the addition of its rows will result in a zero row, implying that the rank cannot exceed $m - 1$. The following theorem demonstrates that the incidence matrix of any orientation of a graph is a representation matrix of the associated cycle matroid.

Theorem 4.2 *Graphic matroids are $\mathbb{R}$-representable.*

Proof For any graph $G(V, E)$, let $A_{\overrightarrow{G}}$ be the incidence matrix of the directed graph $\overrightarrow{G}$ so obtained by applying an arbitrary orientation on G. Similarly as in the proof of Theorem 4.1, we shall prove that $M(G) = M[A_{\overrightarrow{G}}]$ by showing that X is a linearly dependent set of columns in $A_{\overrightarrow{G}}$ if and only if $G[X]$ is contains a cycle in G. Assume that $G[X]$ contains a cycle $G[C]$ for some $C = \{e_1, \dots, e_n\} \subseteq X$. The corresponding columns $\{\mathbf{a}_1, \dots, \mathbf{a}_n\}$ in $A_{\overrightarrow{G}}$ form the submatrix $A_{\overrightarrow{C}}$, which is the incidence matrix of the directed cycle $\overrightarrow{C}$, while each row of $A_{\overrightarrow{C}}$ has exactly two nonzero elements. Choose an arbitrary direction to traverse the cycle, say clockwise, and let $a_i = +1$ if edge e_i has the same direction and $a_i = -1$ otherwise. We have

$$\sum_{i=1}^{n} a_i \mathbf{a}_i = \mathbf{0}. \tag{4.3}$$

The reverse direction is the same as in the proof of Theorem 4.1. □

By a similar argument to the one in the proof of Theorem 4.2 it can be shown that graphic matroids are representable over any field $\mathbb{F}$.

For a matrix A over a field $\mathbb{F}$, we have seen that the sets of linearly independent columns of A constitute the family of independent sets of the vector matroid $M[A]$. It may happen that the representation matrix A contains more information than necessary to define $M[A]$. For example, we may delete any row that is a linear combination of others without altering the matroid. Given an $\mathbb{F}$-representable matroid M with a representation matrix A, the matrix $[\ I\ |\ D\ ]$ obtained from A by applying elementary row operations in $\mathbb{F}$, column interchanges and deletions of zero rows, is called a **standard representation matrix** for M, and the matrix D a **compact representation matrix**. Each standard and compact representation matrix of a matroid is associated with a base of the matroid, as defined by the basis of the matrix. Given a connected directed graph G, the compact representation matrix of $M(G)$ in $\mathbb{R}$ is also known in the literature as the **network matrix** of G and it is equal to $R^{-1}S$, where $[\ R\ |\ S\ ]$ is obtained from the incidence matrix of an arbitrary orientation of G minus an arbitrary row. As it will be demonstrated in Example Sect. 4.1, there is also a combinatorial method of computing the network matrix of a graph.

Example 4.1 Consider the representation matrix in $GF(2)$ for the matroid $M(G)$ of the graph in Fig. 2.1, which by Theorem 4.1 is the matrix A_G given in (4.1). Applying elementary row operations on A_G in $GF(2)$ we obtain a standard representation matrix for $M(G)$ as follows.

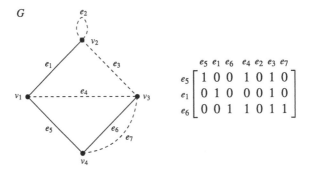

Fig. 4.1 Standard representation matrix of $M(G)$ in $GF(2)$

$$
\begin{array}{c}
\begin{array}{ccccccc} e_1 & e_2 & e_3 & e_4 & e_5 & e_6 & e_7 \end{array}\\
\begin{bmatrix}
1 & 0 & 0 & 1 & 1 & 0 & 0\\
1 & 0 & 1 & 0 & 0 & 0 & 0\\
0 & 0 & 1 & 1 & 0 & 1 & 1\\
0 & 0 & 0 & 0 & 1 & 1 & 1
\end{bmatrix}
\end{array}
\xrightarrow{\begin{array}{l} A_G(1,:) = A_G(1,:) + A_G(2,:)\\ A_G(4,:) = A_G(4,:) + A_G(3,:)\end{array}}
\begin{array}{c}
\begin{array}{ccccccc} e_1 & e_2 & e_3 & e_4 & e_5 & e_6 & e_7 \end{array}\\
\begin{bmatrix}
0 & 0 & 1 & 1 & 1 & 0 & 0\\
1 & 0 & 1 & 0 & 0 & 0 & 0\\
0 & 0 & 1 & 1 & 0 & 1 & 1\\
0 & 0 & 1 & 1 & 1 & 0 & 0
\end{bmatrix}
\end{array}
\xrightarrow{A_G(4,:) = A_G(4,:) + A_G(1,:)}
$$

$$
\begin{array}{c}
\begin{array}{ccccccc} e_1 & e_2 & e_3 & e_4 & e_5 & e_6 & e_7 \end{array}\\
\begin{bmatrix}
0 & 0 & 1 & 1 & 1 & 0 & 0\\
1 & 0 & 1 & 0 & 0 & 0 & 0\\
0 & 0 & 1 & 1 & 0 & 1 & 1\\
0 & 0 & 0 & 0 & 0 & 0 & 0
\end{bmatrix}
\end{array}
\xrightarrow{\begin{array}{l} \text{remove row 4}\\ \text{permute columns}\end{array}}
\begin{array}{c}
\begin{array}{ccc|cccc} e_5 & e_1 & e_6 & e_4 & e_2 & e_3 & e_7 \end{array}\\
\left[\begin{array}{ccc|cccc}
1 & 0 & 0 & 1 & 0 & 1 & 0\\
0 & 1 & 0 & 0 & 0 & 1 & 0\\
0 & 0 & 1 & 1 & 0 & 1 & 1
\end{array}\right]
\end{array}
= [\, I \mid N_{G,T}\,].
$$

The matrix $N_{G,T}$ is the compact representation matrix of $M(G)$ for the basis $T = \{e_5, e_1, e_6\}$. We observe that the columns of $N_{G,T}$ represent characteristic vectors of cycles in G, formed by the inclusion of the corresponding edge into the spanning tree T (see Fig. 4.1). Equivalently, each column e_k is the characteristic vector of the set of elements from the base $T \in \mathcal{B}(M(G))$ that belong to the fundamental circuit $C(e_k, T)$. Pivoting on nonzero elements in $D_{G,T}$ has the effect of replacing elements from the basis. For example, if we want to insert e_4 into the basis, we can see that it can replace any of the elements in $C(e_4, T)$, which are e_5 or e_6. If we want to replace e_5, then we pivot on the $(1, 4)$th element of the standard representation matrix, by adding the first row to the fourth, to obtain

$$
\begin{array}{c}
\begin{array}{ccc|cccc} e_4 & e_1 & e_6 & e_5 & e_2 & e_3 & e_7 \end{array}\\
\begin{array}{c} e_4\\ e_1\\ e_6 \end{array}
\left[\begin{array}{ccc|cccc}
1 & 0 & 0 & 1 & 0 & 1 & 0\\
0 & 1 & 0 & 0 & 0 & 1 & 0\\
0 & 0 & 1 & 1 & 0 & 0 & 1
\end{array}\right]
\end{array},
$$

which is the standard representation matrix that corresponds to the base $\{e_4, e_1, e_6\}$.

If we take the incidence matrix of an orientation $\vec{G}$ of G and apply elementary row operations, we get a standard representation matrix for $M(G)$ in $\mathbb{R}$ as shown

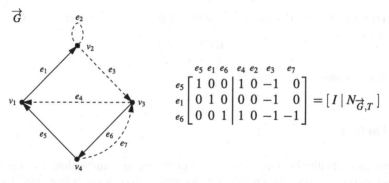

Fig. 4.2 Standard representation matrix of $M(G)$ in $\mathbb{R}$

in Fig. 4.2. The matrix $N_{\vec{G},T}$ is the network matrix of $\vec{G}$ for the spanning tree $T = \{e_5, e_1, e_6\}$, and we have that $N_{\vec{G},T} = N_{G,T} \mod 2$. The elements in a column e of $N_{\vec{G},T}$ can also be obtained by considering the cycle formed by the inclusion of e into the spanning tree T. If we traverse the edges of the cycle in the direction defined by e, we place a $+1$ if an edge has opposite direction and -1 otherwise. $\square$

If we consider the cycles $C_1 = \{e_1, e_3, e_5, e_7\}$ and $C_2 = \{e_4, e_5, e_6\}$ of the graph G in Fig. 2.1, we notice that $C_1 \triangle C_2 = \{e_1, e_3, e_4, e_6, e_7\}$ is the edge set of a disjoint union of cycles of G, while this is the case for any symmetric difference of cycles in the graph. This is a well-known property of graphs, and it is stated in the next theorem.

Theorem 4.3 *If $C_1, C_2, \ldots, C_k$ are cycles in a graph G then the graph $G' = G[E']$ where*

$$E' = C_1 \triangle C_2 \triangle \cdots \triangle C_k,$$

is a disjoint union of cycles of G.

Proof The main observation here is that the degree of every vertex $v \in V(G')$ is even because every symmetric difference operation $C_i \triangle C_{i+1}$ increases $d_{G'}(v)$ by either 0 or 2. We will find a cycle in G'. Consider a $v_0 - v_l$ path in G' of maximum length. Since $d_{G'}(v_0)$ is even v_0 is adjacent to a vertex y other than v_1, which belongs to the $v_0 - v_l$ path, for otherwise the $y - v_l$ path has greater length than l, contradicting our hypothesis. If $y = v_i$ for some $i \in \{2, \ldots, l\}$, we have a cycle C formed by the closed path $v_0, v_1, \ldots, v_i, v_0$. Deleting the edges of C from G' will result in a graph whose vertices have also even degree, therefore the same procedure can be applied. We conclude that $E(G')$ can be partitioned into edge sets of cycles from G. $\square$

We can deduce that the symmetric difference of circuits in a graphic matroid is a disjoint union of circuits. As it is stated in the theorem that follows, this property is not only a necessary condition for graphic matroids, but characterizes the class of binary matroids.

Theorem 4.4 *A matroid M is binary if and only if for any set of circuits $C_1, C_2, \ldots,$
$C_k \in \mathscr{C}(M)$*

$$C_1 \bigtriangleup C_2 \bigtriangleup \cdots \bigtriangleup C_k,$$

is a disjoint union of circuits of M.

4.2 Duality

The notion of duality in matroids is similar to the one in optimization, and it generalizes the concepts of *orthogonality* in vector spaces, and *planarity* in graphs. As the
next theorem demonstrates, for any matroid M we can define another matroid M^* on
the same ground set called the **dual** of M, such that independent sets, bases, circuits,
rank, and any other property of M have well-defined dual counterparts in M^*.

Theorem 4.5 *Given a matroid $M(E, \mathscr{B})$, then*

$$\mathscr{B}^* = \{X \subseteq E : \text{ there exists a base } B \in \mathscr{B}, \text{ such that } X = E - B\}, \qquad (4.4)$$

*is the family of bases of a matroid M^* on E, called the **dual** of M.*

Proof We have to show that the elements of $\mathscr{B}^*$ satisfy axioms (B1) and (B2) of
Theorem 3.2. For (B1), since $\mathscr{B} \neq \emptyset$ there exists some $B \in \mathscr{B}$, therefore $E - B \in \mathscr{B}^*$.

For (B2), consider any two $B_1^*, B_2^* \in \mathscr{B}^*$ and $x \in B_1^* - B_2^*$. We have to show
that there exists $y \in B_2^* - B_1^*$ such that $(B_1^* - \{x\}) \cup \{y\} \in \mathscr{B}^*$, which by (4.4) is
equivalent to the existence of a base $B \in \mathscr{B}$ such that $E - B = (B_1^* - \{x\}) \cup \{y\}$.
Order the elements of E as determined by its partition into $B_1 - B_2$, $B_1 \cap B_2$, $B_2 - B_1$
and $E - (B_1 \cup B_2)$, and set $B_1 = E - B_1^*$ and $B_2 = E - B_2^*$ as it is illustrated in
Fig. 4.3. By (4.4) we have that $B_1, B_2 \in \mathscr{B}$. Moreover, $B_1^* - B_2^* = B_2 - B_1$ therefore
$x \in B_2 - B_1$.

Consider now the set $B_1 \cup \{x\}$. Since $B_1 \in \mathscr{B}$, by Proposition 3.1 $B_1 \cup \{x\}$ will contain a unique circuit $C(x, B_1)$ which will contain x. We also have that $C(x, B_1) - \{x\}$
is not contained in $B_1 \cap B_2$, since $(B_1 \cap B_2) \cup \{x\} \subseteq B_2$. Therefore

$$(C(x, B_1) - \{x\}) \cap (B_1 - B_2) \neq \emptyset,$$

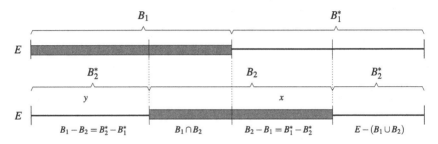

Fig. 4.3 Proof of Theorem 4.5

and for any $y \in C(x, B_1) \cap (B_1 - B_2)$, we have that $(B_1 - \{y\}) \cup \{x\} \in \mathscr{I}$ since it does not contain a circuit. By Lemma 3.1 we have $(B_1 - \{y\}) \cup \{x\} \in \mathscr{B}$ since all bases of M have the same cardinality. Noting that

$$E - (B_1 - \{y\}) \cup \{x\} = (B_1^* - \{x\}) \cup \{y\},$$

axiom (B2) is proved. $\qquad\square$

Clearly, for any matroid M we have $(M^*)^* = M$, since $E - (E - B) = B$ for any $B \in \mathscr{B}$. We will use the prefix *co* whenever referring to a dual notion of a matroid, and an asterisk for the corresponding notation. So we have the dual M^* of the matroid M, which is defined by the families of **coindependent** sets $\mathscr{I}^*(M)$, **cobases** $\mathscr{B}^*(M)$, **cocircuits** $\mathscr{C}^*(M)$, the **corank** function $r^*(M)$, etc. It follows from Theorem 4.5 that

$$r(M) + r^*(M) = |E(M)|.$$

The following theorem establishes the corank of a set.

Theorem 4.6 *For a matroid M and $X \subseteq E(M)$*

$$r^*(X) = |X| - r(M) + r(E - X).$$

Proof Let $B_X^* \in \mathscr{B}^*(X)$ be a maximal coindependent set contained in X, where by Lemma 3.1 we have that $r^*(X) = |B_X^*|$. There exists a cobase $B^* \in \mathscr{B}^*(M)$ that contains B_X^*, and its complement set $B = E - B^*$ is a base of M by Theorem 4.5. The set $B_{E-X} = B \cap (E - X)$ is a maximal independent set contained in $E - X$, which implies that $|B_{E-X}| = r(E - X)$. The situation is depicted in Fig. 4.4. Therefore we have

$$
\begin{aligned}
r^*(X) &= |B_X^*| \\
&= |X| - |X - B_X^*| \\
&= |X| - (|B| - |B_{E-X}|) \\
&= |X| - r(M) + r(E - X).
\end{aligned}
$$

$\qquad\square$

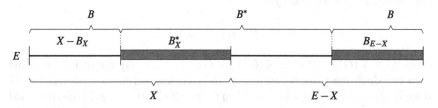

Fig. 4.4 Proof of Theorem 4.6

The following proposition states some immediate relations between a matroid and its dual.

Proposition 4.1 *The following are true for a matroid $M(E, \mathscr{I})$.*

(i) $X \in \mathscr{I}(M)$ *if and only if $E - X$ is cospanning.*
(ii) $X \in \mathscr{H}(M)$ *if and only if $E - X$ is a cocircuit.*

Proof (i) A set $X \subseteq E$ of $M(E, \mathscr{I})$ is spanning if for every $x \in E$ we have $x \in cl(X)$. If $X \in \mathscr{I}(M)$ then there exists a base $B \in \mathscr{B}(M)$ such that $X \subseteq B$. So $E - X$ contains a cobase $B^* = E - B$, and the corank is $r^*(E - X) = r^*(B^*)$. Therefore, for any $x \in E$ we have that $r^*(E - X \cup \{x\}) = r^*(E - X)$, which implies that the set $E - X$ is spanning in M^*. Reversing the argument proves sufficiency. (ii) A set $X \subseteq E$ of $M(E, \mathscr{I})$ is a hyperplane if $cl(X) = X$ and $r(X) = r(E) - 1$. Since for a hyperplane X we have $r(X) < r(E)$ then X is not spanning. But $cl(X) = X$, so for any $y \notin X$ we have $r(X \cup \{y\}) = r(X) + 1 = r(E)$ which implies that $X \cup \{y\}$ is a spanning set. By the dual of (i) we have that $(E - X) \cup \{y\}$ is coindependent, which implies that $E - X$ is a cocircuit. Again, we could state the argument in reverse.   □

Let us now examine duality with respect to the two main classes of matroids in this book, graphic and representable.

Theorem 4.7 *If G is a planar graph with geometric dual G^* then $M^*(G) = M(G^*)$.*

Proof We can assume without loss of generality that G is connected, since on the contrary we can consider each connected component separately. By the definition of the geometric dual we have an one-to-one correspondence between the edges of G and G^*, thus, we can assume that $E(G) = E(G^*)$ and $M^*(G)$ and $M(G^*)$ have a common ground set. We will show that both $M(G)^*$ and $M(G^*)$ have the same family of bases, or equivalently T is a spanning tree of G if and only if $E - T$ is a spanning tree of G^*. Since G is also the geometric dual of G^*, one direction will suffice.

Let $T^* = E - T$. For T^* to be a spanning tree of G^* it must

(i)   contain no cycles,
(ii) contain all vertices of G^*, and
(iii) be connected.

Assume that T^* contains the cycle

$$f_1, e_1, f_2, e_2, \ldots, f_n, e_n, f_1,$$

where $f_i \in V(G^*)$ and $e_i \in E(G^*)$ for $i = 1, \ldots, n$. By the definition of the geometric dual, the pairs of vertices f_i, f_{i+1} correspond to adjacent faces of G which share a common edge $e_i \notin T$. Then the graph $G \backslash \{e_1, \ldots, e_n\}$ is disconnected since $V(G[e_1, \ldots, e_n]) \neq \emptyset$, contradicting the fact that T is a tree in G (see Fig. 4.5). For condition (ii), since T^* does not contain any cycles in G^* we will have

Fig. 4.5 Cycle in G^*

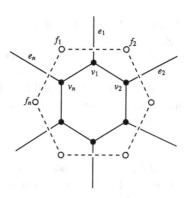

$$|V(T^*)| \geq |E(T^*)| + 1$$
$$= |E(G)| - |E(T)| + 1 \qquad (4.5)$$
$$= |E(G)| - |V(G)| + 2$$
$$= V(G^*),$$

where the last equality is derived from the well-known Euler's formula for the number of faces in a planar graph, since each face of G is a vertex of G^*. Since $T^* \subseteq G^*$ we have $V(T^*) = V(G^*)$. For (iii), since T^* contains no cycles it will be a forest and we have

$$|E(T^*)| = |V(T^*)| + k(T^*),$$

and by (4.5) we have that $k(T^*) = 1$. $\qquad\qquad\qquad\square$

In Sect. 5.1 we will prove that the matroids $M^*(K_{3,3})$ and $M^*(K_5)$ are not graphic matroids. Using this fact, Kuratowski's Theorem which states that a graph is planar if and only if it does not contain as a topological minor K_5 and $K_{3,3}$, and Theorem 4.7, we can prove the following characterization of planar graphs based on the dual of the cycle matroid.

Theorem 4.8 *G is planar if and only if $M(G) = M^*(G)$.*

Therefore, the class of graphic matroids is not dual-closed, and the duals of graphic matroids which are not graphic will be called **cographic**. Those matroids which are both graphic and cographic are called **planar**.

In contrast with the class of graphic matroids, the following theorem states that representable matroids are dual-closed and provides the standard representation matrix for the dual of any vector matroid. It also demonstrates how duality in matroids generalizes the concept of orthogonality in vector spaces.

Theorem 4.9 *If $A = [\, I \mid D \,]$ is a standard representation matrix for a matroid M then $A^* = [\, I \mid -D^T \,]$ is a standard representation matrix for M^*.*

Proof We have to show that $B \in \mathcal{B}(M[A])$ if and only if $(E - B) \in \mathcal{B}(M[A^*])$. Let

$$A = [\ I \mid D \ ], \text{ and } A^* = [\ I \mid -D^T \ ]$$

be matrices in $\mathbb{F}^{r \times n}$. We will first show that the rows of A^* constitute a basis for the nullspace of A. Note that by construction any row of A^* is orthogonal to all the rows of A, that is,

$$A(i, :)A^*(j, :)^T = 0,$$

for all $i \in rows(A)$ and $j \in rows(A^*)$. Therefore, each row of A^* belongs to the nullspace of A. We know from Theorems 1.1 and 2.7 that the dimension of the row space $R(A^T)$ plus the dimension of the nullspace $N(A) = R(A^T)^\perp$ equals to n. Since the dimension of the row space of a matrix equals the dimension of its column space, we have that the dimension of $N(A)$ is $n - r$, therefore the $n - r$ linearly independent rows of A^* constitute a basis for $N(A)$.

Consider now any base $B = \{e_1, \dots, e_r\}$ of $M[A]$, which implies that the corresponding columns form a basis in A. Each column in $\{e_{r+1}, \dots, e_n\}$ in A can be written as a linear combination of the columns in B, so there exist scalars $c_{e_k,i} \in \mathbb{F}$ such that

$$A(:, e_k) + c_{e_k,1} A(:, e_1) + \cdots + c_{e_k,r} A(:, e_r) = \mathbf{0},$$

for each $k = r + 1, \dots, n$. Let the $(n - r) \times n$ matrix formed by these scalars be

$$A' = \begin{matrix} & \begin{matrix} e_1 & e_2 & \cdots & e_r & e_{r+1} & e_{r+2} & \cdots & e_n \end{matrix} \\ & \begin{bmatrix} c_{e_{r+1},1} & c_{e_{r+1},2} & \cdots & c_{e_{r+1},r} & 1 & 0 & \cdots & 0 \\ c_{e_{r+2},1} & c_{e_{r+2},2} & \cdots & c_{e_{r+2},r} & 0 & 1 & \cdots & 0 \\ \vdots & \vdots & \ddots & \vdots & \vdots & \vdots & \ddots & \vdots \\ c_{e_n,1} & c_{e_n,2} & \cdots & c_{e_n,r} & 0 & 0 & \cdots & 1 \end{bmatrix} \end{matrix}.$$

Then the columns with labels $\{e_{r+1}, \dots, e_n\}$ form a basis for A', while since its rows are in $N(A)$, we have that $M[A'] = M[A^*]$ and $E - B$ is a base of $M[A^*]$. The argument can be reversed to show that if $E - B$ is a base of $M[A^*]$ then B is a base of $M[A]$.   □

Example 4.2 Consider the standard representation matrix A_G for the matroid $M(G)$ in Example 4.1. The matrices A_G and A_G^* as well as the graph G and its geometric dual G^* can be seen in Fig. 4.6. We observe that the columns $\{e_4, e_2, e_3, e_7\}$ in A_G^* correspond to a spanning tree of G^*, and the complement of that set to a spanning tree in G. Moreover, each cycle in G is a bond of G^* and vice versa.   □

We get the following corollary to Theorem 4.9.

Corollary 4.1 *A matroid M is binary if and only if M^* is binary.*

Duality in matroids resembles duality in optimization. To see this, consider the discrete optimization problem in (3.14) for a matroid $M(E, \mathcal{B})$. Let

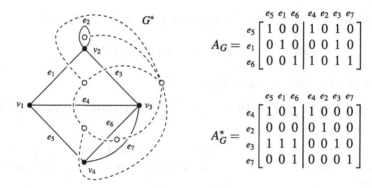

$$A_G = \begin{array}{c} \\ e_5 \\ e_1 \\ e_6 \end{array} \begin{array}{ccc} e_5\ e_1\ e_6 & e_4\ e_2\ e_3\ e_7 \\ \left[\begin{array}{ccc|cccc} 1 & 0 & 0 & 1 & 0 & 1 & 0 \\ 0 & 1 & 0 & 0 & 0 & 1 & 0 \\ 0 & 0 & 1 & 1 & 0 & 1 & 1 \end{array} \right] \end{array}$$

$$A_G^* = \begin{array}{c} \\ e_4 \\ e_2 \\ e_3 \\ e_7 \end{array} \begin{array}{ccc} e_5\ e_1\ e_6 & e_4\ e_2\ e_3\ e_7 \\ \left[\begin{array}{ccc|cccc} 1 & 0 & 1 & 1 & 0 & 0 & 0 \\ 0 & 0 & 0 & 0 & 1 & 0 & 0 \\ 1 & 1 & 1 & 0 & 0 & 1 & 0 \\ 0 & 0 & 1 & 0 & 0 & 0 & 1 \end{array} \right] \end{array}$$

Fig. 4.6 Standard representation matrix of $M^*(G)$

$$B_{\min} = \arg\min\{w(X) \mid X \in \mathcal{B}(M)\},$$
$$B_{\max} = \arg\max\{w(X) \mid X \in \mathcal{B}^*(M)\},$$

be the minimum and maximum weight bases for the M and M^*, respectively. Then we have the relationship

$$B_{\max} = E - B_{\min}.$$

Assuming the contrary, let there exist some cobase $B^* \in \mathcal{B}^*(M)$ such that $w\,(E - B_{\min}) < w(B^*)$. Then $w(B^*) = w(E \backslash B)$ for some $B \in \mathcal{B}(M)$, and since w is a linear function we have $w(E - X) = w(E) - w(X)$ for all $X \subseteq E$, which implies that $w(B_{\min}) > w(B)$, a contradiction.

4.3 Minors

In Sect. 2.1 we have defined the operations of deletion and contraction for graphs. In this section we will show that these operations are dual to each other, and can be generalized to operations in matroids.

Consider for instance the graph $G \backslash \{e_1, e_7\}$ obtained from the graph G of Fig. 2.1. We observe that any cycle of G which has an empty intersection with the set of edges $\{e_1, e_7\}$ remains a cycle in $G \backslash \{e_1, e_7\}$ while all other cycles become acyclic graphs. So we could characterize the set of cycles of $G \backslash \{e_1, e_7\}$ as

$$\{C \subseteq E(G) - \{e_1, e_7\} : C \text{ is a cycle in } G\}.$$

Alternatively, the spanning forests of $G \backslash \{e_1, e_7\}$ are the spanning forests of G which are contained in $E(G) - \{e_1, e_7\}$. The following proposition generalizes the operation of deletion in graphs to matroids.

Proposition 4.2 (Deletion) *For a matroid $M(E, \mathscr{C})$ and $X \subseteq E$, the set*

$$\mathscr{C}(M \backslash X) = \{C \subseteq E - X : C \in \mathscr{C}(M)\}, \tag{4.6}$$

is the family of circuits of a matroid on $E - X$.

Proof The circuit axioms in Theorem 3.3 are trivially satisfied since $\mathscr{C}(M \backslash X) \subseteq \mathscr{C}(M)$.   □

We will call the matroid $M \backslash X$ the **deletion of** X from M. Alternatively, we could define the **deletion to** X in M as the matroid $M|X$ on the ground set X and family of circuits

$$\mathscr{C}(M|X) = \{C \subseteq X : C \in \mathscr{C}(M)\}. \tag{4.7}$$

With respect to the other families that define a matroid as well as the rank and closure, the deletion operation behaves as stated in the next proposition.

Proposition 4.3 *For a matroid M and $X \subseteq E(M)$ we have the following with respect to the operation of deletion.*

  (i) $\mathscr{I}(M \backslash X) = \{Y \subseteq E(M) - X : Y \in \mathscr{I}(M)\}$,
  (ii) $\mathscr{B}(M \backslash X) = maximal\{B - X : B \in \mathscr{B}(M)\}$,
  (iii) $\mathscr{H}(M \backslash X) = maximal\ proper\ subsets\ of\ \{H - X : H \in \mathscr{B}(M)\}$,
  (iv) $r_{M \backslash X}(Y) = r_M(Y)\ for\ all\ Y \subseteq E - X$,
  (v) $cl_{M \backslash X}(Y) = cl_M(Y) - X\ for\ all\ Y \subseteq E - X$.

Proof We have to prove that (i)–(v) are consequences of the family of circuits of $M \backslash X$ given in (4.6). The independence family of $M \backslash X$ in (i) follows from the cryptomorphism between circuits and independent sets given in (3.7). For (ii), since we have $B - X \in \mathscr{I}(M \backslash X)$ for any $B \in \mathscr{B}(M)$, we simply have the definition of the family of bases for $M \backslash X$ as given in Sect. 3.2. Analogously for (iii). For (iv), since $Y \subseteq E \backslash X$ its rank in M is not affected by the deletion of X. Finally for (iv), recall that $cl_M(Y)$ is the set of elements in $E(M)$ that depend on Y, as defined by the rank

$$cl_M(Y) = \{y \in E(M) : r(Y \cup \{y\}) = r(Y)\}.$$

Since the rank of Y is not affected in $M \backslash X$, the result follows.   □

Let us now examine the operation of contraction in graphs. Consider the graph G and its geometric dual G^* as illustrated in Fig. 2.3. If we delete edge e_4 from G, then the graph $G \backslash \{e_4\}$ will remain a planar graph with a geometric dual $(G \backslash \{e_4\})^* = G^*/\{e_4\}$. This is because the deletion of any edge in a planar graph will merge the two neighboring faces into one, which has the effect of identifying the corresponding vertices in the dual graph, or equivalently contracting the edge that connects them. So we can conclude that for a planar graph G and $X \subseteq E(G)$, the operations of contraction and deletion of edges are dual operations, that is,

$$G/X = (G^* \backslash X)^*.$$

Although the above can only be stated for planar graphs, there is no obstacle in extending the concept to the matroid realm.

Definition 4.1 (Contraction) For a matroid M and $X \subseteq E(M)$, the **contraction of X in M** is the matroid M/X on $E(M) - X$ defined as

$$M/X = (M^* \backslash X)^*. \tag{4.8}$$

As in the operation of deletion, we can also define the **contraction to X in M** as the matroid $M.X$ on X defined as

$$M.X = M/(E - X). \tag{4.9}$$

By the definition of the contraction operation in (4.8) and the fact that $(M^*)^* = M$ for a matroid we have the following easy to prove proposition.

Proposition 4.4 For a matroid M and $X \subseteq E(M)$ we have

(i) $(M \backslash X)^* = M^*/X$,
(ii) $(M/X)^* = M^* \backslash X$.

Proof For (ii) take the dual of (4.8), and for (i) do the same for the dual expression of (4.8) which is $M^*/X = (M \backslash X)^*$. □

Let us examine how the family of circuits in a matroid is affected by the operation of contraction.

Proposition 4.5 For a matroid M and $X \subseteq E(M)$

$$\mathscr{C}(M/X) = \text{minimal} \{C - X \ : \ C \in \mathscr{C}(M), C - X \neq \emptyset\}.$$

Proof By the Definition 4.8 we know that the matroid M/X is dual to $M^* \backslash X$, so by (ii) in Proposition 4.1 we have that C is a circuit of M/X if and only if $(E - X) - C$ is a hyperplane of $M^* \backslash X$. By (iii) in Proposition 4.3 we have that $(E - X) - C$ is a maximal proper subset $Y - X$ where $E - Y \in \mathscr{C}(M)$, which is equivalent to C being a minimal nonempty subset $(E - Y) - X$ (see Fig. 4.7). □

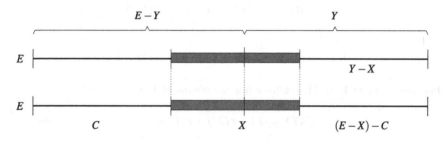

Fig. 4.7 Proof of Proposition 4.5

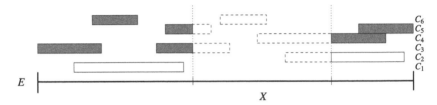

Fig. 4.8 Contraction operation

Example 4.3 Say that we have a matroid M with $\mathscr{C}(M) = \{C_1, C_2, C_3, C_4, C_5, C_6\}$ and some $X \subseteq E(M)$. In Fig. 4.8 the membership of the circuits of M in $E(M)$ is given schematically for an ordering of $E(M)$ such that X appears as depicted. We will have

$$\mathscr{C}(M \backslash X) = \{C_1\}$$

since X meets all circuits except C_1. For the contraction we will have

$$\mathscr{C}(M/X) = \{C_3 - X, C_4 - X, C_5 - X, C_6 - X\}$$

since $C_6 - X \subset C_1$ and $C_4 - X \subset C_2 - X$.   □

In the proof of Proposition 4.5, we described $\mathscr{C}(M/X)$ via $\mathscr{H}(M^* \backslash X)$ using the fact that M/X and $M^* \backslash X$ are dual matroids, and the relationship between circuits and hyperplanes given in Proposition 4.1. Similar argument can be employed for the descriptions of $\mathscr{I}(M/X)$ and $\mathscr{B}(M/X)$ given in the following proposition.

Proposition 4.6 *For a matroid M and $X \subseteq E(M)$ we have the following with respect to the operation of contraction.*

(i) $\mathscr{I}(M/X) = \{Y \subseteq E(M) - X : Y \cup B \in \mathscr{I}(M), B \in \mathscr{B}(M|X)\}$,
(ii) $\mathscr{B}(M/X) = \{Y \subseteq E(M) - X : Y \cup B \in \mathscr{B}(M), B \in \mathscr{B}(M|X)\}$,
(iii) $r_{M/X}(Y) = r_M(X \cup Y) - r_M(X)$ *for all* $Y \subseteq E - X$,
(iv) $cl_{M/X}(Y) = cl_M(X \cup Y) - X$ *for all* $Y \subseteq E - X$.

Proof For (i) and (ii) we follow the same argument as in the proof of Proposition 4.5, where for the dual matroids M/X and $M^* \backslash X$ we have

$$Y \in \mathscr{I}(M/X) \Leftrightarrow (E - X) - Y \in \mathscr{S}(M^* \backslash X),$$

and

$$B \in \mathscr{B}(M/X) \Leftrightarrow (E - X) - B \in \mathscr{B}(M^* \backslash X).$$

For (iii) we know from Theorem 4.6 that the corank of Y is

$$r^*(Y) = |Y| - r(M) + r(E - Y),$$

and $r_{M \backslash X}(Y) = r_M(Y)$. Since M/X and $M^* \backslash X$ are dual matroids we have

$$r_{M/X}(Y) = r_{(M^*\backslash X)^*}(Y)$$
$$= |Y| + r_{M^*\backslash X}(E(M) - X - Y) - r_{M^*\backslash X}(E(M) - X)$$
$$= |Y| + r_{M^*}(E(M) - X - Y) - r_{M^*}(E(M) - X)$$
$$= |Y| + |E(M) - X - Y| + r_M(X \cup Y) - |E(M) - X| - r_M(X)$$
$$= r_M(X \cup Y) - r_M(X).$$

For the closure operator in (iv), we know from Definition 3.10 that

$$cl_{M/X}(Y) = \{y \in E(M) - X : r_{M/X}(Y \cup y) = r_{M/X}(Y)\}. \tag{4.10}$$

From (iii) we have

$$r_{M/X}(Y \cup y) = r_M(Y \cup X \cup y) - r_M(X),$$
$$r_{M/X}(Y) = r_M(Y \cup X) - r_M(X).$$

Replacing the above in (4.10) we get

$$cl_{M/X} = \{y \in E(M) - X : r_M(Y \cup X \cup y) = r_M(Y \cup X)\}$$
$$= \{y \in E(M) : r_M((Y \cup X) \cup y) = r_M(Y \cup X)\} - X$$
$$= cl_M(Y \cup X) - X. \qquad \square$$

The following are basic properties of the operations of deletion and contraction in matroids.

Proposition 4.7 *For a matroid M and disjoint $X, Y \subseteq E(M)$ we have*

(i) $(M\backslash X)\backslash Y = M\backslash(X \cup Y)$.
(ii) $(M/X)/Y = M/(X \cup Y)$.
(iii) $(M/X)\backslash Y = (M\backslash Y)/X$.

Proof We will use the characterizations of the independence family and rank for deletion and contraction, given by Propositions 4.3 and 4.6. For (i) we have

$$\mathscr{I}((M\backslash X)\backslash Y) = \{Z \subseteq E(M) - Y : Z \in \mathscr{I}(M\backslash X)\}$$
$$= \{Z \subseteq E(M) - (X \cup Y) : Z \in \mathscr{I}(M)\}$$
$$= \mathscr{I}(M\backslash(X \cup Y)).$$

For any $Z \subseteq E(M) - (X \cup Y)$, for (ii) we have

$$r_{(M/X)\backslash Y}(Z) = r_{M/X}(Z \cup Y) - r_{M/X}(Y)$$
$$= r_M(Z \cup X \cup Y) - r_M(X \cup Y)$$
$$= r_{M/(X\cup Y)}(Z),$$

and for (iii)

$$r_{(M/X)\backslash Y}(Z) = r_{M/X}(Z)$$
$$= r_M(Z \cup X) - r_M(X)$$
$$= r_{M\backslash Y}(Z \cup X) - r_{M\backslash Y}(X)$$
$$= r_{(M\backslash Y)/X}.$$

□

In the following theorem, we provide some additional properties of the complementary operations of deletion and contraction on non-disjoint sets that will be used in the exposition given in Chap. 5.

Theorem 4.10 *For a matroid* $M(E, \mathscr{C})$ *and* $Y \subseteq X \subseteq E$ *we have*

(i) $(M|X)|Y = M|Y$.
(ii) $(M.X).Y = M.Y$.
(iii) $(M.X)|Y = (M|(E - (X - Y))).Y$.
(iv) $(M|X).Y = (M.(E - (X - Y)))|Y$.

Proof Property (i) follows directly from the Definition 4.6. For property (ii), we will first show that any circuit of $(M.X).Y$ is contained in a circuit of $M.Y$, and vice versa. For any $C \in \mathscr{C}((M.X)/Y$, by (4.8) there exists a circuit $C' \in \mathscr{C}(M.X)$ such that

$$C = C' \cap Y,$$

and for C' there exists a circuit $C'' \in \mathscr{C}(M)$ such that

$$C' = C'' \cap X.$$

Therefore $C = (C'' \cap X) \cap Y$, but since $T \subseteq X$ we have $C = C'' \cap Y$, for $C \neq \emptyset$. Thus, C is contained in a circuit of $M.Y$. Consider now any circuit $C \in \mathscr{C}(M.Y)$. There must exist a circuit $C' \in \mathscr{C}(M)$ such that

$$C = C' \cap Y \subseteq C' \cap X \quad \text{since } Y \subseteq X.$$

Therefore C' is contained in a circuit of $M.X$, say C''. But $C'' \cap Y$ is contained in a circuit $C''' \in \mathscr{C}((M.X)/Y)$ and $C \subseteq C'''$. So for any $X \in \mathscr{C}((M.X).Y)$ there exists $Y \in \mathscr{C}(M.Y)$ such that $X \subseteq Y$. Furthermore, there also exists some $Z \in \mathscr{C}((M.X)/Y)$ such that $Y \subseteq Z$. Therefore

$$X \subseteq Y \subseteq Z,$$

and by axiom (C1) we have $X = Y$, implying that $\mathscr{C}((M.X)/Y) = \mathscr{C}(M.Y)$, which in turn implies that these two matroids are equal.

For property (iii) we will follow a similar approach. Specifically, we will show that any circuit of $(M.X)\backslash Y$ contains a circuit of $(M|(E - (X - Y)).Y)$, and vice versa. Consider any $C \in \mathscr{C}((M.X)|Y)$. Then $C \in \mathscr{C}(M.X)$ such that $C \subseteq Y$, which means that there exists a circuit $C' \in \mathscr{C}(M)$ such that

$$C = C' \cap Y \text{ and } C' \cap (X - Y) = \emptyset.$$

So we have $C' \subseteq E - (X|Y)$, and $C' \in \mathscr{C}(M|(E - (X - Y)))$. Thus, there exists a circuit $C'' \in \mathscr{C}(M|(E - (X - Y)).Y)$ such that $C'' \subseteq C' \cap Y = C$. Now take a circuit $C \in \mathscr{C}(M|(E - (X - Y)).Y)$, which means that there exists a circuit C' of $M|(E - (X - Y))$ such that $C = C' \cap Y$. Therefore, C' is a circuit of M contained in $E - (X - Y)$, so $C' \cap (X - Y) \neq \emptyset$. Since $C' \in \mathscr{C}(M)$, there exists $C'' \in \mathscr{C}(M.X)$ such that

$$C'' \subseteq C' \cap X \subseteq C' \cap Y = C.$$

So we have that $C'' \subseteq Y$ which implies that $C'' \in \mathscr{C}((M.X)|Y)$. Thus, we have that for any $X \in \mathscr{C}((M.X)|Y)$ there exists $Y \in \mathscr{C}(M|(E - (X - Y)).Y)$ such that $Y \subseteq X$, and there exists some $Z \in \mathscr{C}((M.X)|Y)$ such that $Z \subseteq Y$. Therefore by axiom (C1) we have $X = Y$, and the two matroids are equal since they have identical circuit families.

For property (iv), set $X = E - (X - Y)$ in property (iii). □

Given a matroid M there may exist an $X \subseteq E(M)$ such that deletion or contraction of X from M results in the same matroid. For instance, the loop $\{e_4\}$ of graph G in Fig. 2.1 is such a set for the cycle matroid $M(G)$. Examining the definitions of $\mathscr{C}(M \backslash X)$ and $\mathscr{C}(M/X)$ given in Propositions 4.2 and 4.5, respectively, we can deduce that a set $X \subseteq E(M)$ that satisfies $M \backslash X = M/X$ must not *meet* any circuit of M. For instance it is evident that $M \backslash E(M) = M/E(M)$. As it will be demonstrated in Sect. 4.4, minimal sets with the aforementioned property essentially define the connectivity of a given matroid. The following theorem characterizes such sets with respect to their rank.

Theorem 4.11 *For a matroid M and $X \subseteq E(M)$, $M \backslash X = M/X$ if and only if*

$$r(X) + r(E(M) - X) = r(M).$$

Proof ($\Rightarrow$) We know that if $B \in \mathscr{B}(M \backslash X)$ then $r(E(M) - X) = r(B)$, and by assumption $B \in \mathscr{B}(M/X)$. By (ii) in Proposition 4.6 there exists a base $B_X \in \mathscr{B}(M|X)$ such that $B \cup B_X \in \mathscr{B}(M)$. Thus, $r(M) = r(B \cup B_X)$ or equivalently

$$r(M) = r(B) + r(B_X) = r(E(M) - X) + r(X).$$

($\Leftarrow$) Given $\mathscr{I}(M \backslash X)$ and $\mathscr{I}(M/X)$ as defined in Propositions 4.3 and 4.6, respectively, by the independence axiom (I2) we can easily see that $\mathscr{I}(M/X) \subseteq \mathscr{I}(M \backslash X)$. Thus, to prove that the matroids $M \backslash X$ and M/X are equal, it is enough to show that

$\mathcal{I}(M\backslash X) \subseteq \mathcal{I}(M/X)$. For some $Y \in \mathcal{I}(M\backslash X)$, we have that $Y \subseteq B$ for some base $B \in \mathcal{B}(M\backslash X)$, and since $B \in \mathcal{I}(M)$ there is also a base in M which contains B, say $B \cup B'$. So we have

$$r(M) = r(B \cup B') = |B \cup B'| = |B| + |B'| = r(E(M) - X) + |B'|.$$

Since by assumption we have $r(M) = r(X) + r(E(M) - X)$, $|B'| = r(X)$ which implies that $B' \in \mathcal{B}(M|X)$. Using the definition for $\mathcal{B}(M/X)$ in Proposition 4.6 we have that $Y \in \mathcal{I}(M/X)$ since $Y \subseteq B$.   □

Recall that a loop in a matroid is a single element circuit, while a coloop a single element cocircuit. So a loop is not contained in any base, while a coloop is contained in all bases of a matroid. Theorem 4.11 enables us to characterize loops and coloops.

Proposition 4.8 *For a matroid M, an element $e \in E(M)$ is a loop or a coloop if and only if $M\backslash\{e\} = M/\{e\}$.*

Proof Both directions are an immediate consequence of Theorem 4.11.
($\Rightarrow$) If e is a loop then $r(e) = 0$ so $r(e) + r(E(M) - \{e\}) = r(M)$, so by duality $M\backslash\{e\} = M/\{e\}$. If e is a coloop then $r^*(e) = 0$ and, by Theorem 4.6, we will have

$$r^*(e) = |e| - r(M) + r(E(M) - \{e\}) = 0$$
$$\Rightarrow r(e) + r(E(M) - \{e\}) = r(M),$$

since if e is a coloop then $r(e) = |e| = 1$.
($\Leftarrow$) If $M\backslash\{e\} = M/\{e\}$ then $r(e) + r(E(M) - \{e\}) = r(M)$. Assume that e is not a loop. Then $r(e) = 1$ and $r(E(M) - \{e\}) + 1 = r(M)$, which means that $E(M) - \{e\}$ does not contain a basis of M, therefore it is not a spanning set. So by duality the set $E(M) - (E(M) - \{e\}) = e$ is not coindependent, and e is a coloop.   □

In view of Proposition 4.7, the order upon which a sequence of deletions and contractions is applied to a matroid is irrelevant. We can group all elements to be contracted into a set, and all elements to be deleted into another set, the elements in both sets with no specific order. This enables us to define the matroids that are produced by those operations.

Definition 4.2 (**Minor**) For a matroid M and disjoint $X, Y \subseteq E(M)$ the matroid $M\backslash X/Y$ is called a **minor** of M, while if X or Y nonempty it is called **proper**.

A class of matroids $\mathcal{M}$ is called **minor-closed** if every minor of a matroid $M \in \mathcal{M}$ is also a member of the class $\mathcal{M}$. It is easy to find classes of matroids that are not closed under minors. For example, if we define by $\mathcal{M}_n$ the class of matroids with n elements, then both $M\backslash X$ and M/Y are not in $\mathcal{M}_n$ for any nonempty $X, Y \in E(M)$.

Let us examine now the minors of graphic and representable matroids. The next proposition shows that the class of graphic matroids is minor-closed.

Proposition 4.9 *If G is a graph then*

$$M(G)\backslash X/Y = M(G\backslash X/Y)$$

for all $X, Y \subseteq E(G)$.

Proof By Proposition 4.2 and the definition of deletion in graphs given in Sect. 2.1, it follows that

$$\mathscr{C}(M(G)\backslash X) = \mathscr{C}(M(G\backslash X)),$$

for any $X \subseteq E(M)$, hence, these two matroids are equal. Similarly, for the contraction operation, it is easily verified that for any edge $e \subseteq E(G)$ the set

$$\text{minimal } \{C - \{e\} : C \text{ is a cycle of } G\}$$

is the set of cycles of $G/\{e\}$, therefore, by repeated application of (ii) in Proposition 4.7 it follows that

$$\mathscr{C}(M(G)/X) = \mathscr{C}(M(G/X)).$$

$\square$

In order to state an analogous result as that of Proposition 4.9 for the class of $\mathbb{F}$-representable matroids, we have to define the operations of deletion and contraction for matrices. Given a matrix A over a field $\mathbb{F}^{m \times n}$ and some column e, the $m \times (n - 1)$ matrix obtained by deleting column e is called the **deletion of** e, and we write $A\backslash\{e\}$. A series of deletions of columns in $X \subseteq columns(A)$ is denoted by $A\backslash X$. The $(m - 1) \times (n - 1)$ matrix obtained by applying elementary row operations to make column e a unit vector $\mathbf{e}_k$ and deleting column e and row k from A is called the **contraction of** e, and we write $A/\{e\}$. A series of contractions of columns in $Y \in columns(A)$ is denoted by A/Y.

Proposition 4.10 *If A is a matrix over a field* $\mathbb{F}^{m \times n}$ *then*

$$M[A]\backslash X/Y = M[A\backslash X/Y]$$

for all $X, Y \subseteq columns(A)$.

Proof By Proposition 4.2 and the definition of deletion in matrices it follows that

$$\mathscr{I}(M[A]\backslash X) = \mathscr{I}(M[A\backslash X]),$$

for any $X \subseteq columns(A)$. For contraction, by (ii) in Proposition 4.7 it suffices to show that

$$M[A]/\{e\} = M[A/\{e\}]$$

for a column $e \in columns(A)$. If e is a loop it is not contained in any base of M, and we know by Proposition 4.8 that

$$M/\{e\} = M\backslash\{e\}.$$

Assume that $\bar{A}$ is a standard representation matrix for $M[A]$ that corresponds to a base that includes e, that is,

$$\bar{A} = x \begin{array}{cc} X & Y \\ \left[I_n \mid D \right], \end{array}$$

for a partition $\{X, Y\}$ of the columns of A with $e \in X$, where $D \in \mathbb{F}^{n \times (m-n)}$. By Theorem 4.9 we have $M^*[\bar{A}] = M[\bar{A}^*]$ where

$$\bar{A}^* = Y \begin{array}{cc} Y & X \\ \left[I_{m-n} \mid -D^T \right], \end{array}$$

and by the proof for the deletion operation given above we have $M[\bar{A}^*]\backslash\{e\} = M[\bar{A}^*\backslash\{e\}]$, where

$$\bar{A}^*\backslash\{e\} = Y \begin{array}{cc} Y & X-\{e\} \\ \left[I_{m-n} \mid -D^T\backslash\{e\} \right]. \end{array} \tag{4.11}$$

By the definition of the contraction operation on matroids in (4.8), we have that the matroid $M[A]/\{e\} = M[\bar{A}]/\{e\}$ is dual to $M^*[\bar{A}]\backslash\{e\} = M[\bar{A}^*]\backslash\{e\}$, hence its representation matrix is the dual of (4.11), that is,

$$(\bar{A}^*\backslash\{e\})^* = x-\{e\} \begin{array}{cc} X-\{e\} & Y \\ \left[I_{n-1} \mid \bar{D} \right] = A/\{e\}, \end{array}$$

where $\bar{D}$ is the matrix D with row e deleted.   □

Example 4.4 Consider the matroid $M = M[A_G] = M(G)$ for the graph G and the matrix A_G in Example 4.2. We want to compute the standard representation matrix $A_G\backslash\{e_7\}/\{e_3\}$ and the graph $G\backslash\{e_7\}/\{e_3\}$ for the minor $M\backslash\{e_7\}/\{e_3\}$. We will have

$$A_G\backslash\{e_7\} = \begin{array}{ccccccc} & e_5 & e_1 & e_6 & e_4 & e_2 & e_3 \\ & \left[\begin{array}{ccc|ccc} 1 & 0 & 0 & 1 & 0 & 1 \\ 0 & 1 & 0 & 0 & 0 & \boxed{1} \\ 0 & 0 & 1 & 1 & 0 & 1 \end{array} \right] \end{array} \begin{array}{l} A_G(1,:) = A_G(1,:) + A_G(2,:) \\ A_G(3,:) = A_G(3,:) + A_G(2,:) \\ \hline \longrightarrow \end{array} \begin{array}{ccccccc} e_5 & e_1 & e_6 & e_4 & e_2 & e_3 \\ \left[\begin{array}{ccc|ccc} 1 & 1 & 0 & 1 & 0 & 0 \\ 0 & 1 & 0 & 0 & 0 & 1 \\ 0 & 1 & 1 & 1 & 0 & 0 \end{array} \right] \Rightarrow \end{array}$$

$$A_G\backslash\{e_7\}/\{e_3\} = \begin{array}{ccccc} e_5 & e_1 & e_6 & e_4 & e_2 \\ \left[\begin{array}{ccc|cc} 1 & 1 & 0 & 1 & 0 \\ 0 & 1 & 1 & 1 & 0 \end{array} \right] \end{array} = \begin{array}{ccccc} e_5 & e_6 & e_1 & e_4 & e_2 \\ \left[\begin{array}{cc|ccc} 1 & 0 & 1 & 1 & 0 \\ 0 & 1 & 1 & 1 & 0 \end{array} \right] \end{array}$$

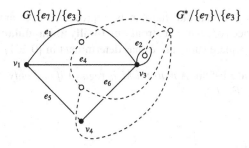

Fig. 4.9 Deletion and contraction operations

It is easily verifiable that the matrix $A_G \setminus \{e_7\}/\{e_3\}$ is a standard representation matrix for the cycle matroid of the graph $G \setminus \{e_7\}/\{e_3\}$ shown in Fig. 4.9. Moreover, the matrix

$$
(A_G \setminus \{e_7\}/\{e_3\})^* = \begin{array}{c} \begin{array}{ccccc} e_1 & e_4 & e_2 & e_5 & e_5 \end{array} \\ \left[\begin{array}{ccc|cc} 1 & 0 & 0 & 1 & 1 \\ 0 & 1 & 0 & 1 & 1 \\ 0 & 0 & 1 & 0 & 0 \end{array} \right] \end{array}
$$

is a standard representation matrix for the cycle matroid of the graph $G^*/\{e_7\}\setminus\{e_3\}$ shown in Fig. 4.9.   □

Classes of matroids which are closed under minors can be characterized by the set of matroids which are not in the class, and are minimal with respect to this property. Specifically, if $\mathscr{M}$ is a minor-closed class of matroids, a matroid $M \notin \mathscr{M}$ is called an **excluded minor** for $\mathscr{M}$ if all its proper minors belong to $\mathscr{M}$. It so happens that the set of excluded minors for major matroid classes are finite. In the following theorems we state excluded minor characterizations for some major classes of matroids.

Theorem 4.12 (Tutte 1958a, b) *A matroid M is binary if and only if it has no minor isomorphic to $U_{2,4}$.*

The set of excluded minors for matroids representable over $GF(3)$ is attributed to Reid who, however, never published his results.

Theorem 4.13 (Bixby 1979; Seymour 1979) *A matroid M is ternary if and only if it has no minor isomorphic to $U_{2,5}$, $U_{3,5}$, F_7, F_7^*.*

The matroid F_7 is the **Fano** matroid, with a compact representation matrix in $GF(2)$

$$
\begin{bmatrix} 1 & 1 & 0 & 1 \\ 1 & 0 & 1 & 1 \\ 0 & 1 & 1 & 1 \end{bmatrix}.
$$

Theorem 4.14 (Tutte 1959) *A matroid M is graphic if and only if it has no minor isomorphic to $U_{2,4}$, F_7, F_7^*, $M^*(K_5)$, and $M^*(K_{3,3})$.*

A **regular** matroid is a matroid that can be represented over any field. The real representation matrices of regular matroids are **totally unimodular**, that is, matrices where each of their square submatrices has determinant in $\{0, \pm1\}$.

Theorem 4.15 (Tutte 1958b) *A matroid M is regular if and only if it has no minor isomorphic to* $U_{2,4}$, F_7, F_7^*.

4.4 Connectivity

Connectivity is a fundamental structural property of matroids, and can be thought of as a measure of correlation between the elements of the ground set with respect to the structure imposed by the family of independent sets, circuits, etc. The more connected a matroid is, the less probable is the existence of sets of elements that are not members of the family that defines the matroid. There are several equivalent ways to define connectivity in matroids. Here, we will adopt the approach by Tutte (1971) as the most appropriate one for the discussion that will follow in Chap. 6 and as the one that more naturally extends to higher connectivity.

Observe that both k-vertex-connectivity and k-edge-connectivity in graphs use the definition of connectivity in graphs, where a path is required between any two vertices for a graph to be connected. Since the notion of a vertex in a matroid is not well-defined, it is not clear how to generalize connectivity to matroids. To exemplify this, consider the two graphs G_1 and G_2 in Fig. 4.10, and their corresponding graphic matroids $M(G_1)$ and $M(G_2)$. We know that the set of cycles in a graph G defines a matroid on $E(G)$, therefore we have that $M(G_1) = M(G_2)$, while G_1 is connected and G_2 is not. Given a matroid $M(E, \mathscr{C})$ assume that there exists a partition $\{X, Y\}$ of E such that if $C \in \mathscr{C}$ then $C \subseteq X$ or $C \subseteq Y$. We could say then that the elements of X are not *related* with the ones in Y with respect to the family $\mathscr{C}$, in the sense that $\mathscr{C}(M\backslash X) \cup \mathscr{C}(M\backslash Y) = \mathscr{C}(M)$. Let us now define the base notion of a separator.

Definition 4.3 For a matroid $M(E, \mathscr{C})$ a set $X \subseteq E$ is called a **separator** of M if any circuit $C \in \mathscr{C}$ is contained in either X or $E - X$.

It follows from the definition of separators that both E and $\emptyset$ are trivial separators for any matroid. Minimal nonempty separators will be called **elementary** separators.

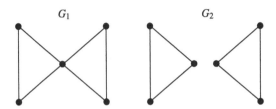

Fig. 4.10 Connectivity in matroids

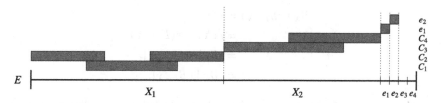

Fig. 4.11 Elementary separators

A matroid will be called **connected** if it has no separators other than the trivial E and $\emptyset$. Note that a singleton $e \in E$ is a separator if and only if e is a loop or there does not exist a circuit that contains e; it is a coloop. To illustrate the aforementioned concepts, consider a matroid $M(E, \mathscr{C})$ where the elements of $E(M)$ are arranged in the horizontal axis as defined by the partition induced by the elementary separators of M, and the members of $\mathscr{C}$ are in the vertical axis as illustrated in Fig. 4.11 for a matroid with six circuits. In this case the elementary separators of M will be $X_1 = C_1 \cup C_2$ and $X_2 = C_3 \cup C_4$, the loops e_1 and e_2, and the coloops e_3 and e_4 which are independent in M and not contained in any circuit. Essentially, the Definition 4.3 states that a separator is a set of elements that does not *meet* any circuit of the matroid, in the sense that a circuit is either contained in the set or is disjoint with the set. It follows that the rank of a separator and the rank of the rest of the elements equals the rank of the matroid. This property is also sufficient to characterize separators.

Proposition 4.11 *For a matroid $M(E, \mathscr{C})$ some set $X \subseteq E$ is a separator of M if and only if $r(X) + r(E - X) = r(E)$.*

Proof ($\Rightarrow$) Assume that X is a separator. By the rank axiom (R3) we have that $r(X) + r(E - X) \geq r(E)$, and let us assume that $r(X) + r(E - X) > r(E)$. There exist bases $B_X \in \mathscr{B}(X)$ and $B_{E-X} \in \mathscr{B}(E - X)$ such that $|B_X| + |B_{E-X}| > r(E)$. Since X is a separator no circuit of M meets both B_X and B_{E-X}, which implies that $B_X \cup B_{E-X}$ contains no circuits, a contradiction since we found an independent set in M with size more than $r(E)$.

($\Leftarrow$) Let $r(X) + r(E - X) = r(E)$ and assume that there exists a circuit C that meets both X and $E - X$, where $C_X = C \cap X$ and $C_{E-X} = C \cap (E - X)$. Since both C_X and C_{E-X} are independent in M, they can be extended to bases $B_X \in \mathscr{B}(X)$ and $B_{E-X} \in \mathscr{B}(E - X)$, respectively. By Definition 3.10 of the closure operator we have that $X \subseteq cl(B_X)$ and $(E - X) \subseteq cl(B_{E-X})$, thus by the closure axiom $(CL2)$ we have

$$E \subseteq cl(B_X) \cup cl(B_{E-X}) \subseteq cl(B_X \cup B_{E-X}).$$

Therefore, $cl(B_X \cup B_{E-X}) = E$ implying that $r(B_X \cup B_{E-X}) = r(E)$. Combining the above we get

$$r(B_X \cup B_{E-X}) = r(E)$$
$$= r(X) + r(E - X)$$
$$= |B_X| + |B_{E-X}|$$
$$= |B_X \cup B_{E-X}|,$$

which means that $B_X \cup B_{E-X}$ is independent, which is a contradiction since it contains the circuit C.   □

We also have the following easy corollary that characterizes separators with respect to the deletion and contraction operations.

Corollary 4.2 *Given a matroid M, a set $X \subseteq E(M)$ is a separator of M if and only if $M \backslash X = M/X$.*

Proof By Proposition 4.2 and Theorem 4.11.   □

In view of Proposition 4.2 we can easily prove that the connectivity of a matroid and its dual coincide.

Proposition 4.12 *Given a matroid M, a set X is separator of M if and only if X is a separator of M^*.*

Proof By Corollary 4.2 and Proposition 4.4, a set X is a separator of M if and only if

$$M \backslash X = M/X \Leftrightarrow$$
$$(M \backslash X)^* = (M/X)^* \Leftrightarrow$$
$$M^*/X = M^* \backslash X,$$

if and only if X is a separator of M^*.   □

Corollary 4.3 *A matroid M is connected if and only if M^* is connected.*

Given a matroid M, let X be a union of circuits $C_1, \ldots, C_k$ of M. Then

$$E - X = E - \bigcup_{i=1}^{k} C_i = \bigcap_{i=1}^{k} (E - C_i) = \bigcap_{i=1}^{k} H_i,$$

where each H_i is a cohyperplane, since if C is a circuit of a matroid then by Proposition 4.1 the set $E - C$ is a cohyperplane. We also know that the intersection of hyperplanes is a flat, therefore $E - X$ has to be a coflat. The following lemma appears in Tutte (1959), and states a useful structural result about unions of circuits, that will be used in the proof of Theorem 5.1.

Lemma 4.1 *Given a matroid M and $Y \in \mathscr{C}(M)$, let X be a union of circuits of $M \backslash Y$ where $M \backslash Y | X$ is connected. Then either $X \cup Y$ is a union of circuits of M such that $M | (X \cup Y)$ connected, or $M/Y | X = M | X$.*

Proof Since any circuit of $M \backslash Y$ is a circuit of M by the definition of the deletion operation, then clearly $X \cup Y$ is a union of circuits of M. If $M|(X \cup Y)$ is connected then there is nothing left to prove. If $M|(X \cup Y)$ is not connected, then we can assume that it has two separators S_1 and S_2. Given that $M \backslash Y|X$ is connected, we must have $S_1 = X$ and $S_2 = Y$. Using now the properties (i) and (ii) of Theorem 4.10 and Corollary 4.2, since $X \subseteq E - Y$ we have

$$
\begin{aligned}
(M/Y)|X &= (M.(E-Y))|X \\
&= (M|(\underbrace{E-(E-Y-X)}_{Y \cup X})).X \\
&= (M|(Y \cup X)).X \\
&= (M|(Y \cup X))|X \\
&= M|X.
\end{aligned}
$$

□

The reason for the multiple definitions for higher connectivity in graphs given in Sect. 2.1, has to do with the various attempts to generalize the concept to matroids. In what follows, we will present two alternative notions of higher connectivity in matroids by quantifying the concept of a separator.

Definition 4.4 (**Matroid k-connectivity**) For a matroid M and a positive integer k, a partition $\{X, Y\}$ of $E(M)$ is a k-**separation** of M if

(i) $\min\{|X|, |Y|\} \geq k$,
(ii) $r(X) + r(Y) \leq r(M) + k - 1$.

The **connectivity number** of matroid M is defined as

$$\lambda(M) = \min\{k : M \text{ has a } k\text{-separation for } k \geq 1\}, \qquad (4.12)$$

while if M does not have a k-separation for any number $k \geq 1$ then $\lambda(M) = \infty$. We say that a matroid M is k-**connected** for any $1 \leq k \leq \lambda(M)$.

If $\{X, E - X\}$ is a 1-separation of a matroid M then by definition we have that $r(X) + r(E - X) - r(M) \leq 0$, and by the submodularity of the rank function $r(X) + r(E - X) - r(M) \geq 0$, which by Proposition 4.11 means that X is a separator of M. So 1-separations are separators, which implies that a matroid is connected if and only if it is 2-connected. As it was the case with matroid connectivity, k-connectivity in matroids is also duality invariant.

Proposition 4.13 *For a matroid M, $\{X, Y\}$ is a k-separation of M if and only if $\{X, Y\}$ is a k-separation of M^*.*

Proof Let $\{X, Y\}$ be a k-separation of M. From Theorem 4.6 and since $r(X) + r(E - X) \leq r(M) + k - 1$ by (ii) in Definition 4.4 we have

$$r^*(X) + r^*(E - X) = |X| - r(M) + r(E - X) + |E - X| - r(M) + r(X)$$
$$= |E| + r(X) + r(E - X) - 2r(M)$$
$$\leq |E| - r(M) + k - 1$$
$$= r^*(M) + k - 1.$$

Letting $\{X, Y\}$ be a k-separation of M^* the above computation also applies for the rank of M since the statement of Theorem 4.6 is self-dual. $\square$

Corollary 4.4 *For a matroid M we have $\lambda(M) = \lambda(M^*)$.*

The next theorem states that k-connectivity in matroids is indeed a generalization of k-connectivity in graphs.

Theorem 4.16 (Tutte 1960) *If G is a connected graph then $\lambda(G) = \lambda(M(G))$.*

From the discussion above we can conclude that the notion of k-connectivity in matroids is duality invariant and generalizes the corresponding notion of k-connectivity in graphs. It is easy to see that a minimal set of k vertices in a graph whose deletion makes the graph disconnected induces a k-separation, therefore, a graph if G is k-connected then it is also k-vertex-connected. The other direction though is not always true, and k-connectivity in matroids does not generalize k-vertex-connectivity in graphs. This prompted for the following alternative notion of higher connectivity in matroids which generalizes vertex connectivity in graphs.

Definition 4.5 (*Matroid vertical k-connectivity*) For a matroid M and a positive integer k, a partition $\{X, Y\}$ of $E(M)$ is a **vertical k-separation** of M if

(i) $\min\{r(X), r(Y)\} \geq k$,
(ii) $r(X) + r(Y) \leq r(M) + k - 1$.

The **vertical connectivity number** of matroid M is defined as

$$\kappa(M) = \min\{k : M \text{ has a vertical } k\text{-separation for } k \geq 1\}, \qquad (4.13)$$

while if M does not have a vertical k-separation for any number $k \geq 1$ then $\kappa(M) = \infty$. We say that a matroid M is **vertical k-connected** for any $1 \leq k \leq \kappa(M)$.

Since $|X| \geq r(X)$ for any $X \subseteq E(M)$ in a matroid M, a vertical k-separation in M induces a k-separation. Hence, if a matroid is k-connected then it is also vertical k-connected. As the next theorem states vertical k-connectivity in matroids generalizes k-vertex-connectivity in graphs, however it is not duality invariant.

Theorem 4.17 (Cunningham 1981; Inukai and Weinberg 1981; Oxley 1981) *If G is a connected graph then $\kappa(G) = \kappa(M(G))$.*

Example 4.5 Consider the graphs G_1, G_1^* and G_2 given in Fig. 4.12. We have that $\kappa(G_1) = 3 = \alpha(G_1)$ while $\lambda(G_1) = 2 = \lambda(M(G_1))$ since we have the 2-separation defined by $\{e_1, e_2\}$. On the other hand, for G_1^* the dual graph of G_1, we have that

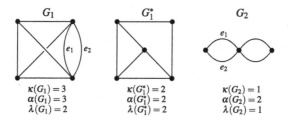

$\kappa(G_1) = 3$
$\alpha(G_1) = 3$
$\lambda(G_1) = 2$

$\kappa(G_1^*) = 2$
$\alpha(G_1^*) = 2$
$\lambda(G_1^*) = 2$

$\kappa(G_2) = 1$
$\alpha(G_2) = 2$
$\lambda(G_2) = 1$

Fig. 4.12 Higher connectivity

$\kappa(G_1^*) = \lambda(G_1^*) = \alpha(G_1^*) = 2$ which shows that the vertex-connectivity and edge-connectivity functions are not duality invariant. For G_2 we have $\kappa(G_2) = 1 = \lambda(G_2)$ and $\alpha(G_2) = 2$. The set $\{e_1, e_2\}$ is a vertical 1-separation of $M(G_2)$, hence $\kappa(M(G_2)) = \kappa(G_2)$ as expected. □

4.5 Notes

Network matrices constitute a well-known class of totally unimodular matrices. Being totally unimodular, network matrices are of importance in optimization since integer programming problems with network constraint matrices have integral polyhedra and they can be solved as linear programs. Moreover, if the constraint matrix of a linear program is a network matrix, then the network simplex algorithm can be used to solve to problem in a more efficient way that the simplex algorithm. In Sect. 5.3 a polynomial time recognition algorithm for network matrices will be presented.

The concept of basic matroid connectivity was included in Whitney's seminal paper (Whitney 1935). Higher connectivity in matroids was introduced by Tutte (1966), where he characterized the class of 3-connected matroids where the deletion or contraction of any element destroys 3-connectivity. Vertical k-connectivity was introduced independently by Cunningham (1981), Oxley (1981), and Inukai and Weinberg (1981), and all three works provided proofs for the Theorem 4.17.

Under certain conditions, the existence of a k-separation in a matroid implies that this matroid can be composed of two smaller matroids which are minors of the original matroid joined together by an operation called **k-sum**, for $k = 1, 2, 3$. One of the most celebrated results in matroid theory is the structural characterization of regular matroids by Seymour (1980), where it is proved that any regular matroid can be decomposed via k-sums into graphic and cographic matroids and one special matroid with ten elements.

Chapter 5
Decomposition of Graphic Matroids

In the series of papers (Tutte 1956, 1958a, b, 1959) Tutte developed a rich theory regarding the structure of regular, binary, and graphic matroids, and provided among other results the excluded minor characterizations stated in Theorems 4.12 and 4.14. Consider the following problem, which will serve as our primary motivation for the material presented in this chapter.

Problem 5.1 (Graph Realization) Given a matrix A in $GF(2)$ find a graph G such that $M[A] = M(G)$, or decide that no such graph exists.

The graph realization problem is equivalent to the question of whether a binary matroid given by a representation matrix is graphic. The characterization of graphic matroids given in Theorem 4.14 does not lead to an algorithm for solving the graph realization problem, since there is no efficient way of checking whether a matroid contains a given minor or not.[1]

Apart from the excluded minor characterization for graphic matroids, Tutte (1959) also provided a decomposition theorem for graphic matroids which, as it is often with decomposition results, leads to an efficient recursive recognition algorithm that solves the graph realization problem (Tutte 1960). A decomposition result for a matroid class typically consists of a set of well-defined decomposition and composition operations on minors, and a theorem that proves that the class is closed under these operations. Usually the direction which proves that the composition of minors from a class results in a matroid within the class is the most difficult. There are relatively few characterizations of matroid classes based on decomposition results, the most notable one being the decomposition of regular matroids by Seymour (1980). In this chapter we will present the theory related to the decomposition of graphic matroids, and give an algorithm that solves the graph realization problem which is a direct consequence of that theory.

---

[1] Although recent work on Matroid Minor Theory by J. Geelen, B. Gerards and G. Whittle may provide a polynomial time algorithm for checking whether a matroid representable over a finite field contains a given minor.

L. S. Pitsoulis, *Topics in Matroid Theory*,
SpringerBriefs in Optimization, DOI: 10.1007/978-1-4614-8957-3_5,
© Leonidas S. Pitsoulis 2014

5.1 Bridges

Let Y be a cocircuit of a binary matroid M. We define the **bridges** of Y in M to be the elementary separators of $M \backslash Y$. If $M \backslash Y$ has more than one bridge then we say that Y is a **separating** cocircuit; otherwise it is **non-separating**. Let B be a bridge of Y in M; the matroid $M.(B \cup Y)$ is called a Y**-component** of M. The following is a useful result regarding the connectivity of the Y-components.

Theorem 5.1 *If M is a connected matroid then each Y-component of M is connected.*

Proof By duality it is enough to show that $M^*|(B \cup Y)$ is connected. Consider any bridge B of Y in M. Then B is also an elementary separator of M^*/Y, since a matroid is connected if and only if its dual is connected. Observe that B cannot be an element $\{e\} \in \mathcal{I}(M^*.Y)$, that is, a coloop of $M^*.Y$. This would imply that $\{e\}$ is a loop of $(M^*.Y)^* = M \backslash Y$ which means that it is a loop of M and a coloop of M^*, thus, M^* is not connected contradicting the fact that M is connected. Therefore, we can assume that the elementary separator B is a union of circuits of $M^*.Y$, such that $(M^*.Y)|B$ is connected. By Lemma 4.1, either $M^*|(B \cup Y)$ is connected, or

$$(M^*/Y)|B = M^*|B.$$

Since B is an elementary separator of M^*/Y, we have

$$\begin{aligned} M^*|B &= (M^*/Y)|B \\ &= (M^*/Y).B \\ &= M^*.(E - Y).B \\ &= M^*.B \end{aligned}$$

by property (ii) of Theorem 4.10 since $B \subseteq E - Y$. Therefore, B is a separator of M^* which is a contradiction.   □

For any bridge B of Y in M, we denote by $\pi(M, B, Y)$ the family of all minimal non-null subsets of Y which are intersections of cocircuits of $M.(B \cup Y)$. The following theorem and its corollary relate $\pi(M, B, Y)$ for binary matroids with the family of cocircuits of a given minor.

Theorem 5.2 *Let Y be a cocircuit of a matroid M. Two elements a and b of Y belong to the same members of $\pi(M, B, Y)$ if and only if they belong to the same cocircuits of $M.(B \cup Y)|Y$.*

Proof ($\Leftarrow$) Suppose that $a, b \in W \in \pi(M, B, Y)$. Then for any cocircuit X of $M.(B \cup Y)$ either $X \cap W = \emptyset$ or $W \subseteq X$, since otherwise W will not be minimal. This implies that a and b belong to exactly the same cocircuits of $M.(B \cup Y)$. Thus, by the definition of the matroid operations of contraction and deletion we have that a and b belong to the same cocircuits of $M.(B \cup Y)|Y$.

($\Rightarrow$) Since a and b belong to the same cocircuits of $M.(B \cup Y)|Y$ then by the definition of matroid contraction and deletion we obtain that there is no cocircuit Z of $M.(B \cup Y)$ such that $Z \cap \{a, b\} = \{a\}$ or $Z \cap \{a, b\} = \{b\}$. Therefore, by the definition of the members of $\pi(M, B, Y)$, the result follows. $\square$

Tutte (1965) proved that if M is binary, then the members of $\pi(M, B, Y)$ are disjoint and their union is Y. We usually refer to $\pi(M, B, Y)$ as the partition of Y determined by B. By this result, Theorem 5.2 has the following useful corollary.

Corollary 5.1 *Let Y be a cocircuit of a matroid M. If M is binary then*

$$\pi(M, B, Y) = \mathscr{C}^*(M.(B \cup Y)|Y).$$

It can be shown that for a given graph G, a set of edges X is an elementary separator of $M(G)$ if an only if

(i) $G|X$ is 2-connected,
(ii) each connected component of $G[E(G) - X]$ has at most one vertex in common with $G|X$.

Let us call any subgraph of G which satisfies properties (i) and (ii) a **separate** of G. Therefore, the separates of a graph correspond to the elementary separators of its cycle matroid, and vice versa. So if $\{B_1, \ldots, B_k\}$ is the set of bridges of Y in $M(G)$, then each bridge B_i will appear as a separate $G|B_i$ in one of the end-graphs of $G \backslash Y$. Moreover, given the structure of the separates in the graph and the fact that each component is connected, all the bridges of the same component will form a tree-like structure in the sense that any path between a pair of vertices from distinct bridges will pass through a unique set of bridges. Note that by Corollary 5.1 and the fact that the cocircuits of a graphic matroid are bonds in the associated graph, we have a graphical characterization of $\pi(M(G), B, Y)$ for some cocircuit Y and bridge B. Suppose now that B is a bridge of Y in $M(G)$ and let G_i be the component of $G \backslash Y$ such that $G|B \subseteq G_i$. Then, if v is a vertex of $V(G|B)$, we denote by $C(B, v)$ the component of $G_i \backslash B$ having v as a vertex. Moreover, we denote by $Y(B, v)$ the set of all $y \in Y$ such that one end of y in G is a vertex of $C(B, v)$. The following provides an alternative direct characterization in G of the partition of Y as determined by B, in a graphic matroid $M(G)$

Theorem 5.3 *Let $M(G)$ be a graphic matroid and $Y \in \mathscr{C}^*(M(G))$. If $G|B$ is a separate of an end-graph G_i of $G \backslash Y$, then $\pi(M(G), B, Y)$ is the class of all nonempty $Y(B, v)$ such that $v \in V(G|B)$.*

Proof Let G_1 and G_2 be the end-graphs of the bond Y in G. We know from Corollary 5.1 and Proposition 4.9 that

$$\pi(M(G), B, Y) = \mathscr{C}^*(M(G).(B \cup Y)|Y)$$
$$= \mathscr{C}^*(M(G.(B \cup Y)|Y)).$$

Thus, it is enough to show that the set of bonds of $G.(B \cup Y)|Y$ is

$$\mathscr{L} = \{Y(B, v) \neq \emptyset : v \in V(G|B) \}.$$

Assume that $G|B$ is a subgraph of G_1. The graph $G.(B \cup Y)$ is obtained from G by contracting G_2 into a single vertex u, and $C(B, v)$ into a vertex v, for every $v \in V(G|B)$. Therefore, in the graph $G.(B \cup Y)|Y$ each vertex $v \neq u$ is connected only to u by the set of parallel edges $Y(B, v)$, which means that $Y(B, v)$ is a bond in $G.(B \cup Y)|Y$. It remains to show that $G.(B \cup Y)|Y$ contains no other bonds. By Corollary 5.1, the cocircuits of $\mathscr{C}^*(M(G.(B \cup Y)|Y))$, which are the bonds of $G.(B \cup Y)|Y$, are disjoint. Given that $\mathscr{L}$ is a partition of the edges of $G.(B \cup Y)|Y$, the existence of a bond not in $\mathscr{L}$ would imply the existence of two bonds which are not disjoint, a contradiction.   □

The following example illustrates the definitions given so far.

Example 5.1 We will illustrate how bridges and Y-components manifest as separates in graph representations of graphic matroids. Let $M(G)$ be the cycle matroid of the graph G in Fig. 5.1. If $Y = \{y_1, y_2, y_3, y_4\} \in \mathscr{C}^*(M(G))$ then Y is a bond of G, and the bridges of Y are the separates of $G \backslash Y$, that is, B_1, B_2 and B_3. The corresponding Y-components are the graphs $G.(B_i \cup Y)$, while the partitions of Y by its bridges $\pi(M, B_i, Y)$ are the sets of bonds of the graphs $G.(B_i \cup Y)|Y$, for $i = 1, 2, 3$. Moreover, we have

$$C(B_1, v_1) = G[\{v_1, v_4\}]$$
$$C(B_1, v_2) = G[\{v_2\}]$$
$$C(B_1, v_3) = G[\{v_3\}]$$

resulting in $Y(B_1, v_1) = \{y_3, y_4\}$, $Y(B_1, v_2) = \{y_1, y_2\}$ while $Y(B_1, v_3) = \emptyset$. Therefore, $\pi(M, B_1, Y) = \{Y(B_1, v_1), Y(B_1, v_2)\}$. Similarly for $\pi(M, B_2, Y)$ and $\pi(M, B_3, Y)$.   □

Let us see now how the bridges of a cocircuit in a graphic matroid relate to each other, with respect to the associated graph. Assume that B_1 and B_2 are two bridges of Y in M. The bridges B_1 and B_2 are said to **avoid** each other if there exist $S \in \pi(M, B_1, Y)$ and $T \in \pi(M, B_2, Y)$ such that $S \cup T = Y$; otherwise we say that B_1 and B_2 **overlap** one another. A cocircuit Y is called **bridge-separable** if its bridges can be partitioned into two classes U and V such that all members of the same class avoid each other.

Example 5.2 For the cocircuit Y of Example 5.1, we have

$$\pi(M, B_1, Y) = \{\{y_1, y_2\}, \{y_3, y_4\}\},$$
$$\pi(M, B_2, Y) = \{\{y_1, y_2, y_3\}, \{y_4\}\},$$
$$\pi(M, B_3, Y) = \{\{y_1\}, \{y_3\}, \{y_2, y_4\}\},$$

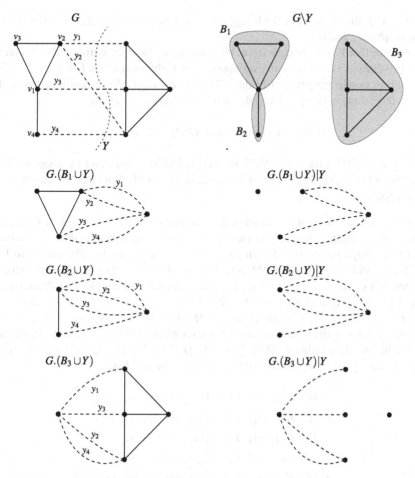

Fig. 5.1 Bridges and Y-components

therefore, B_2 avoids B_1 and B_3, but B_1 overlaps with B_3. So we have that Y is bridge-separable, since any bipartition of its bridges with B_1 and B_3 in different classes will suffice. We can also observe that $C(B_1, v_1) = B_2$ and $C(B_2, v_1) = B_1$, since we only have two 2-connected components in the end graph that results from the deletion of Y from G. □

It turns out that bridge-separability of cocircuits is a necessary condition for graphic matroids.

Theorem 5.4 *If M is a graphic matroid and $Y \in \mathscr{C}^*(M)$ then Y is bridge-separable.*

Proof Let the graph G be such that $M = M(G)$, and G_1, G_2 be the end-graphs of $G \setminus Y$. Denote with

$$\mathscr{L}_i = \{B \subseteq E(M) : G|B \text{ is a subgraph of } G_i\},$$

for $i = 1, 2$, the partition of the bridges of Y as defined by their membership in the end-graphs G_1 and G_2.

Assume that Y is not bridge-separable. This means that any partition of its bridges into two classes will contain an overlapping pair in the same class. Let $B_1, B_2 \in \mathcal{L}_1$ be such an overlapping pair of bridges. Then $G|B_1$ and $G|B_2$ are both separates of $G \setminus Y$, thus, there exist $v_1 \in V(G|B_1)$ and $v_2 \in V(G|B_2)$ such that

$$G|B_1 \subseteq C(B_2, v_2) \text{ and } G|B_2 \subseteq C(B_1, v_1).$$

Now if $v \in V(G_1)$ then $v \in V(C(B_1, v_1)) \cup V(C(B_2, v_2))$, and by Theorem 5.3 we have $Y(B_1, v_1) \cup Y(B_2, v_2) = Y$, which is a contradiction since B_1 and B_2 are overlapping. $\square$

Let us apply Theorem 5.4, to show that the matroids $M^*(K_5)$ and $M^*(K_{3,3})$ are not graphic matroids. Note that in view of Theorem 4.7, this implies the well-known fact from graph theory that K_5 and $K_{3,3}$ are not planar graphs. The cocircuits of $M^*(K_5)$ will be the circuits of $M(K_5)$, which will be the sets of edges that induce cycles in K_5. Let $Y = \{y_1, y_2, y_3, y_4, y_5\}$ be a cocircuit of $M^*(K_5)$, as depicted in Fig. 5.2. The bridges of Y in $M^*(K_5)$ will be the separates of $(K_5/Y)^*$, which are $G|\{e_i\}$ for $i = 1, \ldots, 5$. To determine $\pi(M^*(K_5), \{e_i\}, Y)$ for each bridge e_i, by Corollary 5.1 it suffices to consider the cocircuits of $M^*(K_5).(Y \cup \{e_i\})|Y$, which by duality are the circuits of $M(K_5)|(Y \cup \{e_i\}).Y = M(K_5|(Y \cup \{e_i\}).Y)$, which in turn are the cycles of the graph $K_5|(Y \cup \{e_i\}).Y$. We will have

$$\pi(M^*(K_5), \{e_1\}, Y) = \{\{y_1, y_5\}, \{y_2, y_3, y_4\}\},$$
$$\pi(M^*(K_5), \{e_2\}, Y) = \{\{y_2, y_3\}, \{y_1, y_4, y_5\}\},$$
$$\pi(M^*(K_5), \{e_3\}, Y) = \{\{y_1, y_2\}, \{y_3, y_4, y_5\}\},$$
$$\pi(M^*(K_5), \{e_4\}, Y) = \{\{y_4, y_5\}, \{y_1, y_2, y_3\}\},$$
$$\pi(M^*(K_5), \{e_5\}, Y) = \{\{y_3, y_4\}, \{y_1, y_2, y_5\}\}.$$

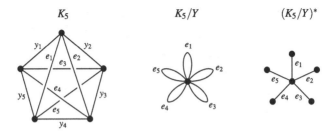

Fig. 5.2 Overlapping bridges in $M^*(K_5)$

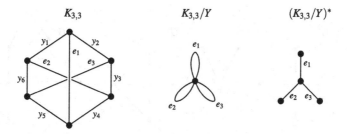

Fig. 5.3 Overlapping bridges in $M^*(K_{3,3})$

Upon inspection we observe that all bridges of Y overlap with one another, therefore, by Theorem 5.4 the matroid $M^*(K_5)$ cannot be graphic.

Consider now $K_{3,3}$ as depicted in Fig. 5.3. For $Y = \{y_1, y_2, y_3, y_4, y_5, y_6\}$ a cocircuit of $M^*(K_{3,3})$, the bridges of Y will be $\{\{e_1\}, \{e_2\}, \{e_3\}\}$ and

$$\pi(M^*(K_{3,3}), \{e_1\}, Y) = \{\{y_1, y_5, y_6\}, \{y_2, y_3, y_4\}\},$$
$$\pi(M^*(K_{3,3}), \{e_2\}, Y) = \{\{y_1, y_2, y_3\}, \{y_4, y_5, y_6\}\},$$
$$\pi(M^*(K_{3,3}), \{e_3\}, Y) = \{\{y_3, y_4, y_5\}, \{y_1, y_2, y_6\}\}.$$

As previously, we can see that any pair of bridges is overlapping, and $M^*(K_{3,3})$ is not a graphic matroid.

If a matroid has a cocircuit which is not bridge-separable, then it will contain a minor isomorphic to $M^*(K_5)$, $M^*(K_{3,3})$ or F_7^*.

Instead of cocircuits with overlapping bridges consider now cocircuits that have bridges which avoid each other. The graphical significance of all avoiding bridges in a cocircuit of a graphic matroid is illustrated in Example 5.3.

Example 5.3 For the graph G in Fig. 5.1, take the graph $G' = G \backslash \{y_2\}$, and the bond $Y' = \{y_1, y_3, y_4\}$ in this graph, as shown in Fig. 5.4. In this case Y' has the same bridges B_1, B_2, and B_3 in G' as Y in G, but now we have

$$\pi(M, B_1, Y') = \{\{y_1\}, \{y_3, y_4\}\},$$
$$\pi(M, B_2, Y') = \{\{y_1, y_3\}, \{y_4\}\},$$
$$\pi(M, B_3, Y') = \{\{y_1\}, \{y_3\}, \{y_4\}\},$$

where one can check that all bridges of Y avoid each other. Note that we can find 2-separations in G' such that after two consecutive twistings about the defining vertices $\{v, w\}$ we can get G'''. The bond Y is now a star in G''', which has the same set of cycles as G', so we have that $M(G''') = M(G')$.   □

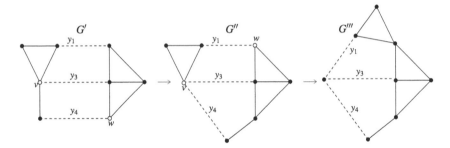

Fig. 5.4 All avoiding bridges

In Theorem 5.5 we prove that if a cocircuit of a graphic matroid has all avoiding bridges, then there is a graph representation of the matroid where this cocircuit is a star of a vertex. This result is vital for the proof of the composition part of the Theorem 5.6, since it provides the link between graphic matroids and their representations as graphs. Specifically, given that a cocircuit Y of a graphic matroid is bridge-separable, then its bridges can be arranged into two distinct classes of all avoiding bridges, and we know that the corresponding matroids have graph representations with Y being a star. A simple composition operation then can be defined where the stars of the two graphs identify into a bond of a union of the graphs.

Theorem 5.5 *Let Y be a cocircuit of a connected graphic matroid M such that any two bridges of Y avoid each other. Then there exists a 2-connected graph G where Y is a star of a vertex, and $M = M(G)$.*

Proof There exists a 2-connected graph G such that $M(G) = M$, and Y is a bond in G. Let G_1, G_2 be the two components of $G \backslash Y$. We shall show that there exist disjoint 2-separations in G, such that, by a series of twistings on the associated vertices of the 2-separations we can reduce the size of G_1 by one separate at a time, until we are left with a single vertex whose star will be Y.

Fix an arbitrary bridge B_0 of Y in $M(G)$ where $G|B_0$ is a separate of G_2. For any bridge B_1 of Y such that $G|B_1$ is a separate of G_1, we know, by Theorem 5.3 and the fact that all bridges of Y are avoiding, that there exist a pair of vertices $v_0 \in V(G|B_0)$ and $v_1 \in V(G|B_1)$ such that

$$Y(B_0, v_0) \cup Y(B_1, v_1) = Y. \tag{5.1}$$

Choose B_1, v_1 and v_0 such that the number of edges of $C(B_1, v_1)$ is the least possible. For a bridge B and $v \in V(G|B)$, if G_i is the component of $G \backslash Y$ such that $G|B \subseteq G_i$, define $F(B, v) = G_i \backslash E(C(B, v))$. Let $B_1, B_2, \ldots, B_k$ be the bridges of Y that contain vertex v_1, and consider any one of them, say B_i. The situation for $k = 4$ is depicted in Fig. 5.5. We know that there exist $w \in V(G|B_i)$ and $v \in V(G|B_0)$ such that

$$Y(B_i, w) \cup Y(B_0, v) = Y. \tag{5.2}$$

Fig. 5.5 Groups of bridges as 2-separations in G

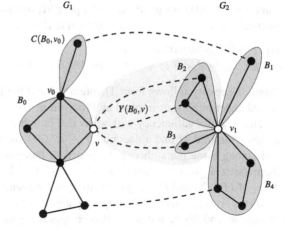

We will show that $w = v_1$ for all $i = 1, \ldots, k$. From (5.1) and the fact that G is 2-connected, we can deduce that there exists at least one edge $e \in Y$ with one end-vertex in $F(B_1, v_1)$ and the other end-vertex in $C(B_0, v_0)$. Suppose now that $w \neq v_1$. Then $e \notin Y(B_i, w)$ which implies that $v = v_0$ for (5.2) to be true. This contradicts the choice of B_1, v_1 and v_0 since $E(C(B_i, w)) \subset E(C(B_1, v_1))$.

Let $Q(v)$ be the group of bridges B_i such that (5.2) holds (with B_0 fixed). For example, in Fig. 5.5, both B_2 and B_3 satisfy (5.2) with B_0. Each such group $Q(v)$ defines a 2-separation in G with defining vertices v and v_1 and a partition $\{T, E(G) \backslash T\}$ of $E(G)$, where

$$T = E(C(B_0, v)) \cup Y(B_0, v) \cup \left(\bigcup_{B_i \in Q(v)} E(F(B_i, v_1)) \right),$$

since for any $B_i \in Q(v)$ for $i = 1, \ldots, k$, there are no edges of Y with one end-vertex in $F(B_i, v_1)$ and the other in $F(B_0, v)$ due to the avoidance between B_i and B_0. Therefore, by twisting about v_1 and v for every group of bridges $Q(v)$, we can create a graph G' where Y is a bond such that one of the components of $G' \backslash Y$ is G_2 with B_0 replaced by a vertex, while the other component is G_1 with v_1 replaced by B_0. □

5.2 Decomposition

The main result of this chapter is the following theorem, which in contrast with Theorem 4.14, provides a structural characterization of graphic matroids.

Theorem 5.6 (Decomposition of Graphic Matroids) *Let M be a connected binary matroid and $Y \in \mathscr{C}^*(M)$. M is graphic if and only if*

(i) *Y is bridge-separable, and*
(ii) *for any bridge B of Y, the minor $M.(B \cup Y)$ is graphic.*

Proof Necessity follows from Theorem 5.4 and the fact that graphic matroids are closed under minors, as implied by Proposition 4.9.

The proof for sufficiency will be divided into three parts. Assume that there exist a binary matroid M and a cocircuit $Y \in \mathscr{C}^*(M)$ such that the theorem is not true, and among those pairs choose the one with the least $|E(M)|$. We will prove that there exists a graph G such that $M = M(G)$. If Y has only one bridge B, then $M = M.(B \cup Y)$ and M is a graphic matroid by assumption. So assume that Y has at least two bridges. Since Y is bridge-separable, we can partition its bridges into two classes $\mathscr{U}_1$ and $\mathscr{U}_2$ such that any two bridges in the same class avoid each other. Let

$$E_i = \bigcup_{B \in \mathscr{U}_i} B, \quad i = 1, 2,$$

be the elements in $E(M)$ of the members in each class. For the rest of the proof assume that $i = 1, 2$.

Part 1 *The matroids $M.(E_1 \cup Y)$ and $M.(E_2 \cup Y)$ are graphic matroids.* Assume that $M.(E_i \cup Y)$ is not connected, and has a separator $S \subseteq E_i \cup Y$. Then, there must exist a bridge $B \in \mathscr{U}_i$ of Y in M such that $S \cap (B \cup Y) \neq \emptyset$. By the definition of the contraction operation, $S \cap (B \cup Y)$ would also be a separator of $M.(B \cup Y)$ which is a contradiction since, by Theorem 5.1, we know that every Y-component is connected. Thus, $M.(E_i \cup Y)$ is connected. The cocircuits of $M.(E_i \cup Y)$ are the circuits of $(M.(E_i \cup Y))^* = M^*|(E_i \cup Y)$, which are the cocircuits of M contained in $E_i \cup Y$, so Y is a cocircuit of $M.(E_i \cup Y)$. For any bridge $B \in \mathscr{U}_i$ of Y in M, since B is a separator of $M \backslash Y$, then it is also a separator of $(M.(E_i \cup Y)) \backslash Y$. Therefore, we have that B is a bridge of Y in $M.(E_i \cup Y)$ and

$$\pi(M.(E_i \cup Y), B, Y) = \pi(M, B, Y),$$

which implies that any two bridges of Y in $M.(E_i \cup Y)$ avoid each other; therefore, Y is bridge-separable. Finally, the corresponding Y-component of Y in $M.(E_i \cup Y)$ for the bridge B is

$$(M.(E_i \cup Y)).(B \cup Y) = M.(B \cup Y),$$

by property (ii) of Theorem 4.10, and this matroid is a graphic matroid. We conclude that $M.(E_i \cup Y)$ satisfies the conditions of the theorem, and since it has less elements than M it is a graphic matroid by assumption.

Part 2 *Construction of a graph G.* Since all the bridges of Y in $\mathscr{U}_i$ avoid each other, by Theorem 5.5 there exists a 2-connected graph G_i such that Y is a star of a vertex w_i and

Fig. 5.6 Proof of Theorem 5.6

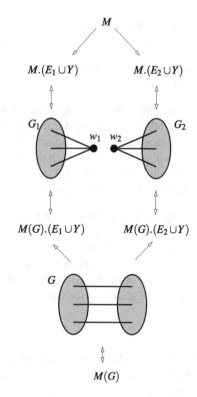

$$M.(E_i \cup Y) = M(G_i).$$ (5.3)

Construct a graph G with $E(G) = E(M)$ by adjoining $G_1 \backslash Y$ and $G_2 \backslash Y$ as follows: For any edge $e \in Y$, add an edge between the end-vertex of e in G_1 to the corresponding end-vertex of e in G_2 and name that edge e (see Fig. 5.6). Since the graphs G_1 and G_2 are 2-connected and Y is a star of a vertex in both, $G \backslash Y$ will have two connected components by construction, which implies that Y is a bond of G and a cocircuit in the cycle matroid $M(G)$. We can now relate M and $M(G)$ through $M(G_i)$, since by Proposition 4.9 and (5.3) we have

$$M(G).(E_i \cup Y) = M(G.(E_i \cup Y)) = M(G_i) = M.(E_i \cup Y).$$ (5.4)

Part 3 *Prove that $M = M(G)$.* In this part we will make use of the characterization of binary matroids given in Theorem 4.4, which can be stated for the cocircuits of a binary matroid due to Corollary 4.1. Having established a relationship between M and $M(G)$ as given by (5.4), we will now employ a matroid argument to prove that these two matroids are, in fact, equal. Consider the following family of cocircuits of $M(G)$

$$\mathcal{K} = \{X \in \mathcal{C}^*(M(G)) : \exists\, X_i \in \mathcal{C}^*(M) \text{ such that } X = \triangle X_i\}.$$

Note that for $X_1, X_2 \in \mathcal{K}$ such that $X_1 \cap X_2 \neq \emptyset$, since $M(G)$ is binary, we have that $X_1 \triangle X_2 \in \mathcal{C}^*(M(G))$ and $X_1 \triangle X_2 \in \mathcal{K}$.

Claim   There exists some $X \in \mathcal{C}^*(M(G)) - \mathcal{K}$ such that $X - Y$ is a cocircuit of $M(G) \backslash Y$.

Proof   We can assume that for all $X \in \mathcal{C}^*(M(G)) - \mathcal{K}$ we have $X \cap Y = \emptyset$, since otherwise by the deletion operation we have that $X - Y \in \mathcal{C}^*(M(G) \backslash Y)$. Choose such an X and assume that it is not a cocircuit of $M(G) \backslash Y$. Then there exists $T \in \mathcal{C}^*(M(G))$ such that $T \subset X \cup Y$ and, since $M(G)$ is binary $X \triangle T$ is a cocircuit of $M(G)$. If $X \triangle T$ or T do not belong to $\mathcal{K}$ the result follows. Therefore, $(X \triangle T) \triangle T = X$ which implies that $X \in \mathcal{K}$, which is a contradiction.

By the above claim and the fact that E_1 and E_2 are separators for $M(G) \backslash Y$ by construction, we can conclude that either $X \subseteq (E_1 \cup Y)$ or $X \subseteq (E_2 \cup Y)$. Therefore, since $X \in \mathcal{C}^*(M(G))$, we have that X is either a cocircuit of $M(G).(E_1 \cup Y) = M.(E_1 \cup Y)$ or a cocircuit of $M(G).(E_2 \cup Y) = M.(E_2 \cup Y)$. In both cases X is a cocircuit of M. But since M is connected and binary this is a contradiction to the fact that $X \notin \mathcal{K}$. So for any cocircuit $X \in \mathcal{C}^*(M(G))$ we have

$$X = X_1 \triangle X_2 \triangle \ldots \triangle X_n$$

for $X_i \in \mathcal{C}^*(M)$. But by the dual of Theorem 4.4 in binary matroids the symmetric difference of cocircuits contains a cocircuit, or it is empty. So, we can conclude that there exists some $X' \in \mathcal{C}^*(M)$ such that $X' \subseteq X$.

Reversing the above argument we can also conclude that any cocircuit X' of M contains a cocircuit X of $M(G)$, and by Lemma 3.2 we have $M = M(G)$, a contradiction to our original hypothesis.   □

The argument for proving sufficiency in the above proof can be seen schematically in Fig. 5.6. For a binary matroid M two smaller matroids $M.(E_1 \cup Y)$ and $M.(E_2 \cup Y)$ are constructed and it is proved that they are graphic matroids and Y is a cocircuit with all-avoiding bridges in both. Therefore, by Theorem 5.5 we can find two graphs G_1 and G_2 where Y is a star of a vertex in both, thus, we can compose a larger graph G and prove that $M = M(G)$.

The above theorem can be viewed as a decomposition of graphic matroids in the sense that if some well-defined minors of a matroid are graphic then the matroid is also graphic, and vice versa. In other words, the property of graphicness is not only inherited to the minors of the matroid, but even more importantly, it is maintained in the composition of the matroid from these minors. A direct consequence of this, is a recognition algorithm, that we will discuss in the next section.

5.3 Recognition Algorithm for Graphic Matroids

Theorem 5.6 can be used to construct a polynomial-time recognition algorithm which determines whether a binary matroid given by a $\{0, 1\}$ representation matrix is graphic and provide the corresponding graph in the affirmative. We will first provide a sketch of the algorithm and then describe it in full detail. Given a binary matroid M, first we have to find a cocircuit Y such that $M\backslash Y$ is not connected. As we will see later on, if M is graphic then there is always a separating cocircuit which is easy to compute. For each separator B of $M\backslash Y$, if the collection of cocircuits of the minor $M.(B \cup Y)|Y$ does not partition Y, then the matroid is not graphic. Equivalently, if we cannot partition the bridges of Y into two classes of all-avoiding bridges, the matroid is not graphic. If the aforementioned conditions are satisfied, we apply the same procedure to each minor $M.(B \cup Y)$ for any bridge B of Y. Eventually, the size of these minors will become small enough such that they will be graphic matroids.

Before describing the recognition algorithm in detail, we have to define some preliminary operations and procedures that will be used by the algorithm as subroutines. Although Theorem 5.6 is stated for matroids and graphs, the recognition algorithm description will be restricted to matrices representing matroids and graphs. All matrices as well as any operations on them are defined on the binary field $GF(2)$. Consider two incidence matrices R_{G_1} and R_{G_2}, and let Y be the star of a vertex in both G_1 and G_2. Let

$$
R_{G_1} = \begin{array}{c} V_1 \\ w_1 \end{array}\left[\begin{array}{c|c} E_1 & Y \\ \hline R_{11} & R_{12} \\ \hline 0 \cdots 0 & 1 \cdots 1 \end{array}\right], \quad R_{G_2} = \begin{array}{c} V_2 \\ w_2 \end{array}\left[\begin{array}{c|c} E_2 & Y \\ \hline R_{21} & R_{22} \\ \hline 0 \cdots 0 & 1 \cdots 1 \end{array}\right],
$$

where E_1, E_2, and V_1, V_2 are the index sets of the columns and rows, respectively, and w_1, w_2 are the rows of the matrices R_{G_1} and R_{G_2} that correspond to the characteristic vector of Y. The incidence matrix of the **star composition** of G_1 and G_2 in Y, is defined as

$$
R_{G_1} \circledast_Y R_{G_2} = \begin{array}{c} V_1 \\ V_2 \end{array}\left[\begin{array}{c|c|c} E_1 & E_2 & Y \\ \hline R_{11} & \mathbf{0} & R_{12} \\ \hline \mathbf{0} & R_{21} & R_{22} \end{array}\right],
$$

where $\mathbf{0}$ denotes the matrix of all zeros of appropriate size. The star composition defined above is the equivalent matrix operation to the one described for graphs in part 2 of the proof of Theorem 5.6. In the following, assume that we have an incidence matrix R of a graph G. The procedure STANDARD(R) returns the standard representation matrix $[I\,|\,R']$, by performing elementary row operations, permutations of the columns, and deletion of any zero row or column in R. For a set of columns Y, the procedure STAR(R, Y) returns the incidence matrix of a graph where Y is the star of a vertex by applying elementary row operations on R, provided that such a representation exists.

The algorithm is shown in pseudocode in Algorithm 5.1. As an input to the algorithm GRAPHIC we give a binary matrix R, and the algorithm returns the incidence matrix R_G of a graph G such that $M(G) = M[R]$, if the matroid $M[R]$ is graphic. If not, the algorithm terminates without any output. The algorithm is recursive in nature, and it terminates if one of the conditions upon which $M[R]$ is not graphic is satisfied in lines 11, 20, and 23, or if R is trivially the incident matrix of a graph.

In lines 1–6, we initially transform R into a standard representative matrix for $M[R]$ and check to see if each column contains at most two nonzeros. In the affirmative, we adjoin the mod 2 sum of the rows in line 4, thereby, constructing a binary matrix with exactly two nonzeros in each column which is the incidence matrix of a graph by definition. Note that the operation in line 4 does not affect linear independence in $GF(2)$, which implies that $M[R] = M[R_G]$.

Algorithm 5.1 GRAPHIC

Input: matrix R in $GF(2)$
Output: incidence matrix R_G of graph G such that $M(G) = M[R]$

  $R := \text{STANDARD}(R)$
  if each column of R has at most two nonzeros $\rightarrow$
    $\mathbf{r} := \sum_{i \in rows(R)} R(i, :)$
    $R_G := \left[\dfrac{R}{\mathbf{r}} \right]$ #$M[R] = M[R_G]$ is graphic
    return R_G
  end if
  Choose a column j and rows i_1, i_2, i_3 such that $R(i_k, j) = 1, \ k = 1, 2, 3$
  for $k = 1, \ldots, 3 \rightarrow$
    $Y := \{j : R(i_k, j) = 1\}$
    $\{B_1, \ldots, B_n\} := \text{SEPARATORS}(R \backslash Y)$
    if $n = 1$ and $k = 3 \rightarrow$ **exit**   # is Y separating?
    if $n > 1 \rightarrow$ **continue**
  end for
  for each $B_k \in \{B_1, \ldots, B_n\} \rightarrow$
    $R_k := \text{STANDARD}(R.(B_k \cup Y)|Y)$
    for each $i \in rows(R_k) \rightarrow$
      $C_i := \{j : R_k(i, j) = 1\}$
      $\pi(B_k) \leftarrow C_i$
    end for
    if $\pi(B_k)$ is not a partition of $Y \rightarrow$ **exit**   # does B_k partitions Y?
  end for
  $\{U_1, U_2\} := \text{AVOIDING}(\{\pi(B_1), \ldots, \pi(B_n)\})$
  if $U_1 = \emptyset$ and $U_2 = \emptyset \rightarrow$ **exit**   # is Y bridge-separable?
  for $k = 1, 2 \rightarrow$
    for each $B_k \in U_i \rightarrow$
      $R_k := \text{GRAPHIC}(R.(B_k \cup Y))$ # is $M[R].(B_k \cup Y)$ graphic?
      $R_k := \text{STAR}(R_k, Y)$
      $R_{U_i} := \text{STAR}(R_{U_i} \circledast_Y R_k, Y)$
    end for
  end for
  $R_G := R_{U_1} \circledast_Y R_{U_2}$
  return R_G

In the case where R has a column j with at least three nonzero elements, from the rows in R that have a nonzero element in column j, one of them has to be a separating cocircuit for $M[R]$ to be graphic. This is so, since a non-separating cocircuit in a graphic matroid corresponds to a star in any graphical representation and an edge of a graph cannot be a member of three distinct stars. This is examined in lines 8–13, where for each row with a nonzero element of column j we compute the separators of $M[R]\backslash Y$, where Y is the cocircuit defined by that row. The procedure SEPARATORS partitions the matrix $R\backslash Y$, and returns the partition of the columns $\{B_1, \ldots, B_n\}$. This can be done efficiently in a number of ways, one possibility is the following. Pick an arbitrary column j and add it to a set B_1. Now add to B_1 any column that has a nonzero element in the rows that have nonzero element in column j, and repeat this procedure for any column in B_1 that has not been examined yet. In the same fashion construct $B_2, \ldots, B_n$ until all columns of $R\backslash Y$ have been included. If at any time during that process we find two or more separators, then we exit the loop in line 12, while if all cocircuits are non-separating, then in line 11 we conclude that the matroid is not graphic and terminate the algorithm without any output.

At this point we have a separating cocircuit Y of $M[R]$, as well as a set $\{B_1, \ldots, B_n\}$ of bridges of Y, where $n > 1$. The next task is to check whether each B_k partitions Y, that is, to check whether the members of $\pi(M[R], B_k, Y)$ are disjoint and their union is Y. By Corollary 5.1, we know that $\pi(M[R], B_k, Y)$ is the collection of cocircuits of $M[R].(B_k \cup Y)|Y$. Therefore, it is enough to check whether each column in a standard form of $R.(B_k \cup Y)|Y$ has exactly one nonzero entry, which is done in lines 16–20. The procedure AVOIDING in line 22, partitions the bridges $\{B_1, \ldots, B_n\}$ into two disjoint classes U_1 and U_2, such that any two bridges in a class are avoiding. This can be done as follows. Given $\pi(M[R], B_i, Y)$ and $\pi(M[R], B_j, Y)$ for any two bridges B_i and B_j, it is trivial to check whether they avoid each other. Construct a graph with vertices $\{B_1, \ldots, B_k\}$, and join any two vertices by an edge if the corresponding pair of bridges is overlapping. If the resulting graph is bipartite, which can be checked easily, then the procedure AVOIDING returns the vertex partitions as U_1 and U_2. If not, then Y is not bridge-separable and the procedure returns $U_1 = U_2 = \emptyset$, where in this case the algorithm terminates in line 23.

The algorithm now has to check whether the Y-components $M_k = M[R].(B_k \cup Y)$ for each bridge B_k are graphic matroids, and if they are, compose the graph G from the graphs G_k. This is done in lines 24–30. In line 26, we get the incident matrix for the graph G_k such that $M(G_k) = M[R.(B_k \cup Y)]$ if the corresponding Y-component is graphic. If not, then the algorithm terminates with no output. By Theorem 5.5 we know that the binary matroid $M[R.(\cup_{B \in U_1} B \cup Y)]$ has a 2-connected graphical representation where Y is a star of a vertex. The incidence matrix of this graph is constructed in line 28, by taking the star compositions one pair at a time. The fact that star representations of the matrices R_k and R_{U_i} in lines 27 and 28 exist, is guaranteed by Theorem 5.5. Similarly for the bridges of U_2. Finally, the incidence matrix of G is computed by the star composition of the resulting graphs in line 31, and returned by the algorithm.

Note that in view of Theorem 4.7 algorithm GRAPHIC can also be used to test whether a given graph G is planar or not.

5.4 Numerical Example

In this section we will go through a numerical example step-by-step in order to get a better understanding of the Algorithm 5.1. Consider that we are given the following standard representation matrix in $GF(2)$

$$
R = \begin{array}{c} \\ 1 \\ 2 \\ 3 \\ 4 \\ 5 \\ 6 \\ 7 \end{array}
\begin{array}{c}
\begin{array}{ccccccccccccccc} e_1 & e_2 & e_3 & e_4 & e_5 & e_6 & e_7 & e_8 & e_9 & e_{10} & e_{11} & e_{12} & e_{13} & e_{14} \end{array} \\
\left[\begin{array}{ccccccc|ccccccc}
1 & 0 & 0 & 0 & 0 & 0 & 0 & 1 & 1 & 1 & 1 & 1 & 0 & 0 \\
0 & 1 & 0 & 0 & 0 & 0 & 0 & 1 & 0 & 1 & 0 & 0 & 0 & 1 \\
0 & 0 & 1 & 0 & 0 & 0 & 0 & 0 & 1 & 0 & 1 & 0 & 0 & 1 \\
0 & 0 & 0 & 1 & 0 & 0 & 0 & 1 & 0 & 1 & 0 & 0 & 0 & 0 \\
0 & 0 & 0 & 0 & 1 & 0 & 0 & 1 & 1 & 1 & 1 & 1 & 0 & 0 \\
0 & 0 & 0 & 0 & 0 & 1 & 0 & 0 & 0 & 1 & 1 & 1 & 1 & 0 \\
0 & 0 & 0 & 0 & 0 & 0 & 1 & 0 & 1 & 1 & 1 & 1 & 1 & 0
\end{array}\right]
\end{array}, \qquad (5.5)
$$

and let $M = M[R]$. We want to find a graph G such that $M = M(G)$, or decide conclusively that no such graph exists. Each row in R is the characteristic vector of a cocircuit of M. Choose a row from R that has a nonzero in a column with three or more nonzero elements, say row 1. This corresponds to the cocircuit $Y_1 = \{e_1, e_8, e_9, e_{10}, e_{11}, e_{12}\}$. The representation matrix for the matroid $M \backslash Y_1$ is obtained from R by deleting the columns that correspond to the elements of Y_1

$$
R \backslash Y_1 = \begin{array}{c} \\ 2 \\ 3 \\ 4 \\ 5 \\ 6 \\ 7 \end{array}
\begin{array}{c}
\begin{array}{cccccccc} e_2 & e_3 & e_4 & e_5 & e_6 & e_7 & e_{13} & e_{14} \end{array} \\
\left[\begin{array}{cccccc|cc}
1 & 0 & 0 & 0 & 0 & 0 & 0 & 1 \\
0 & 1 & 0 & 0 & 0 & 0 & 0 & 1 \\
0 & 0 & 1 & 0 & 0 & 0 & 0 & 0 \\
0 & 0 & 0 & 1 & 0 & 0 & 0 & 0 \\
0 & 0 & 0 & 0 & 1 & 0 & 1 & 0 \\
0 & 0 & 0 & 0 & 0 & 1 & 1 & 0
\end{array}\right]
\end{array},
$$

and its columns and rows can be partitioned as

$$
R \backslash Y_1 = \begin{array}{c} \\ 2 \\ 3 \\ 4 \\ 5 \\ 6 \\ 7 \end{array}
\begin{array}{c}
\begin{array}{cccccccc} e_2 & e_3 & e_{14} & e_4 & e_5 & e_6 & e_7 & e_{13} \end{array} \\
\left[\begin{array}{cccccccc}
1 & 0 & 1 & 0 & 0 & 0 & 0 & 0 \\
0 & 1 & 1 & 0 & 0 & 0 & 0 & 0 \\
0 & 0 & 0 & 1 & 0 & 0 & 0 & 0 \\
0 & 0 & 0 & 0 & 1 & 0 & 0 & 0 \\
0 & 0 & 0 & 0 & 0 & 1 & 0 & 1 \\
0 & 0 & 0 & 0 & 0 & 0 & 1 & 1
\end{array}\right]
\end{array}.
$$

Therefore, we can conclude that Y_1 is a separating cocircuit, and its bridges in M are

$$B_1 = \{e_2, e_3, e_{14}\}, \quad B_2 = \{e_4\}, \quad B_3 = \{e_5\}, \quad B_4 = \{e_6, e_7, e_{13}\}.$$

The Y_1-components $M.(B_i \cup Y_1)$ for $i = 1, \ldots, 4$ are given by the following representation matrices:

$$R.(B_1 \cup Y_1) = \begin{array}{c} 1 \\ 2 \\ 3 \end{array}\left[\begin{array}{ccc|cccccc} 1 & 0 & 0 & 1 & 1 & 1 & 1 & 1 & 0 \\ 0 & 1 & 0 & 1 & 0 & 1 & 0 & 0 & 1 \\ 0 & 0 & 1 & 0 & 1 & 0 & 1 & 0 & 1 \end{array}\right], \tag{5.6}$$

$$\begin{array}{cccccccc} & e_1 & e_2 & e_3 & e_8 & e_9 & e_{10} & e_{11} & e_{12} & e_{14} \end{array}$$

$$R.(B_2 \cup Y_1) = \begin{array}{c} 1 \\ 4 \end{array}\left[\begin{array}{cc|cccccc} 1 & 0 & 1 & 1 & 1 & 1 & 1 \\ 0 & 1 & 1 & 0 & 1 & 0 & 0 \end{array}\right], \tag{5.7}$$

$$R.(B_3 \cup Y_1) = \begin{array}{c} 1 \\ 5 \end{array}\left[\begin{array}{cc|cccccc} 1 & 0 & 1 & 1 & 1 & 1 & 1 \\ 0 & 1 & 1 & 1 & 1 & 1 & 1 \end{array}\right], \tag{5.8}$$

$$R.(B_4 \cup Y_1) = \begin{array}{c} 1 \\ 6 \\ 7 \end{array}\left[\begin{array}{ccc|cccccc} 1 & 0 & 0 & 1 & 1 & 1 & 1 & 1 & 0 \\ 0 & 1 & 0 & 0 & 0 & 0 & 1 & 1 & 1 \\ 0 & 0 & 1 & 0 & 1 & 1 & 1 & 1 & 1 \end{array}\right]. \tag{5.9}$$

Note that $R.(B_1 \cup Y_1) = R/\{e_4, e_5, e_6, e_7, e_{13}\}$, so in order to contract the elements $\{e_4, e_5, e_6, e_7\}$, since these are basic elements, we simply delete the corresponding rows. The columns of the non-basic elements, $\{e_{13}\}$ in this case, will become zero and they can be simply deleted since contraction of a loop in a matroid is equal to deletion. Similarly for the other Y_1-components.

Now we have to compute $\pi(B_i, M, Y_1) = \mathscr{C}^*(M.(B_i \cup Y_1)|Y_1)$ for each bridge B_i of Y_1. The standard representation matrix for $M.(B_1 \cup Y_1)|Y_1$ is obtained from the matrix in (5.6) by deleting the columns that correspond to the elements of B_1, and performing elementary row operations. We will have

$$R.(B_1 \cup Y_1)|Y_1 = \begin{array}{c} 1 \\ 2 \\ 3 \end{array}\left[\begin{array}{cccccc} 1 & 1 & 1 & 1 & 1 & 1 \\ 0 & 1 & 0 & 1 & 0 & 0 \\ 0 & 0 & 1 & 0 & 1 & 0 \end{array}\right] \rightarrow \begin{array}{c} 1 \\ 2 \\ 3 \end{array}\left[\begin{array}{cccccc} 1 & 0 & 0 & 0 & 0 & 1 \\ 0 & 1 & 0 & 1 & 0 & 0 \\ 0 & 0 & 1 & 0 & 1 & 0 \end{array}\right],$$

where we added rows 2 and 3 to row 1. Since all the columns in the standard representation matrix of $M.(B_1 \cup Y_1)|Y_1$ have only one nonzero element, it means that the cocircuits defined by the rows of $R.(B_1 \cup Y_1)|Y_1$ are the only cocircuits of this matroid and they partition the ground set Y_1. Therefore, we have

$$\pi(B_1, M, Y_1) = \{\{e_1, e_{12}\}, \{e_9, e_{11}\}, \{e_8, e_{10}\}\}.$$

Similarly for B_2, B_3 and B_4 we compute

$$R.(B_2 \cup Y_1)|Y_1 = \begin{array}{c} 1 \\ 4 \end{array} \begin{matrix} e_1 & e_8 & e_9 & e_{10} & e_{11} & e_{12} \\ \left[\begin{matrix} 1 & 1 & 1 & 1 & 1 & 1 \\ 0 & 1 & 0 & 1 & 0 & 0 \end{matrix}\right] \end{matrix} \rightarrow \begin{array}{c} 1 \\ 4 \end{array} \begin{matrix} e_1 & e_8 & e_9 & e_{10} & e_{11} & e_{12} \\ \left[\begin{matrix} 1 & 0 & 1 & 0 & 1 & 1 \\ 0 & 1 & 0 & 1 & 0 & 0 \end{matrix}\right] \end{matrix},$$

with $\pi(B_2, M, Y_1) = \{\{e_1, e_9, e_{11}, e_{12}\}, \{e_8, e_{10}\}\}$,

$$R.(B_3 \cup Y_1)|Y_1 = \begin{array}{c} 1 \\ 5 \end{array} \begin{matrix} e_1 & e_8 & e_9 & e_{10} & e_{11} & e_{12} \\ \left[\begin{matrix} 1 & 1 & 1 & 1 & 1 & 1 \\ 0 & 1 & 1 & 1 & 1 & 1 \end{matrix}\right] \end{matrix} \rightarrow \begin{array}{c} 1 \\ 5 \end{array} \begin{matrix} e_1 & e_8 & e_9 & e_{10} & e_{11} & e_{12} \\ \left[\begin{matrix} 1 & 0 & 0 & 0 & 0 & 0 \\ 0 & 1 & 1 & 1 & 1 & 1 \end{matrix}\right] \end{matrix},$$

with $\pi(B_3, M, Y_1) = \{\{e_1\}, \{e_8, e_9, e_{10}, e_{11}, e_{12}\}\}$, and

$$R.(B_4 \cup Y_1) = \begin{array}{c} 1 \\ 6 \\ 7 \end{array} \begin{matrix} e_1 & e_8 & e_9 & e_{10} & e_{11} & e_{12} \\ \left[\begin{matrix} 1 & 1 & 1 & 1 & 1 & 1 \\ 0 & 0 & 0 & 0 & 1 & 1 \\ 0 & 0 & 1 & 1 & 1 & 1 \end{matrix}\right] \end{matrix} \rightarrow \begin{array}{c} 1 \\ 6 \\ 7 \end{array} \begin{matrix} e_1 & e_8 & e_9 & e_{10} & e_{11} & e_{12} \\ \left[\begin{matrix} 1 & 1 & 0 & 0 & 0 & 0 \\ 0 & 0 & 0 & 0 & 1 & 1 \\ 0 & 0 & 1 & 1 & 0 & 0 \end{matrix}\right] \end{matrix},$$

with $\pi(B_4, M, Y_1) = \{\{e_1, e_8\}, \{e_9, e_{10}\}, \{e_{11}, e_{12}\}\}$. We can conclude that each bridge B_1, B_2, B_3, and B_4 partitions Y_1. Moreover, we can see that B_1 and B_2 overlap with B_4 while any other pair of bridges is avoiding. Thus, the collections $U_1 = \{B_1, B_2, B_3\}$ and $U_2 = \{B_4\}$ consist of avoiding bridges, which means that Y_1 is bridge-separable.

We now have to see if the Y_1-components are graphic matroids. Let $M_1 = M.(B_1 \cup Y_1)$ and $R_1 = R.(B_1 \cup Y_1)$. Each column of R_1 has at most two nonzero entries, therefore, by adjoining an extra row that is the mod 2 sum of its rows, it will become the incidence matrix of a graph, while the matroid will remain the same. Let R_1' be such a matrix, where we apply an arbitrary labeling in its rows and row w is the mod 2 sum of the rows of R_1

$$R_1' = \begin{array}{c} v_1 \\ v_2 \\ v_3 \\ w_1 \end{array} \begin{matrix} e_2 & e_3 & e_{14} & e_1 & e_8 & e_9 & e_{10} & e_{11} & e_{12} \\ \left[\begin{array}{ccc|cccccc} 1 & 0 & 1 & 0 & 1 & 0 & 1 & 0 & 0 \\ 0 & 1 & 1 & 0 & 0 & 1 & 0 & 1 & 0 \\ 1 & 1 & 0 & 1 & 0 & 0 & 0 & 0 & 1 \\ \hline 0 & 0 & 0 & 1 & 1 & 1 & 1 & 1 & 1 \end{array}\right] \end{matrix}.$$

Note that $M_1 = M[R_1] = M[R_1']$ and Y_1 will also be a cocircuit of M_1. The graph G_1 that corresponds to the incidence matrix R_1' is shown in Fig. 5.7, where the bond Y_1 is indicated. It is also clear from the graph that the cocircuits of $M.(B_1 \cup Y_1)|Y_1$ which are the bonds of $G_1|Y_1$, are given by $\pi(B_1, M, Y_1)$.

If we let $M_2 = M.(B_2 \cup Y_1)$, $M_3 = M.(B_3 \cup Y_1)$ and $R_2 = R.(B_2 \cup Y_1)$, $R_3 = R.(B_3 \cup Y_1)$, we observe that the matrices R_2 and R_3 also have at most two

G_1

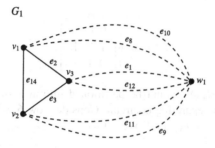

Fig. 5.7 Graph G_1 such that $M(G_1) = M.(B_1 \cup Y_1)$

nonzero elements in each column. Applying the same procedure as we did for the matrix R_1 we get the incidence matrices

$$R_2' = \begin{array}{c} \\ v_4 \\ v_5 \\ w_2 \end{array} \begin{array}{cc} \begin{array}{ccccccc} e_4 & e_1 & e_8 & e_9 & e_{10} & e_{11} & e_{12} \end{array} \\ \left[\begin{array}{c|cccccc} 1 & 0 & 1 & 0 & 1 & 0 & 0 \\ 1 & 1 & 0 & 1 & 0 & 1 & 1 \\ 0 & 1 & 1 & 1 & 1 & 1 & 1 \end{array} \right] \end{array}$$

$$R_3' = \begin{array}{c} \\ v_6 \\ v_7 \\ w_3 \end{array} \begin{array}{cc} \begin{array}{ccccccc} e_5 & e_1 & e_8 & e_9 & e_{10} & e_{11} & e_{12} \end{array} \\ \left[\begin{array}{c|cccccc} 1 & 0 & 1 & 1 & 1 & 1 & 1 \\ 1 & 1 & 0 & 0 & 0 & 0 & 0 \\ 0 & 1 & 1 & 1 & 1 & 1 & 1 \end{array} \right] \end{array}$$

of the graphs G_2 and G_3 shown in Fig. 5.8.

Thus far we have concluded that the Y_1-components M_1, M_2, and M_3 are graphic matroids, and we have found corresponding graphical representations G_1, G_2, and G_3. It remains to examine $M_4 = M.(B_4 \cup Y_1)$. Letting $R_4 = R.(B_4 \cup Y_1)$ in (5.9), we observe that R_4 has columns with more than three nonzero elements, hence we cannot readily conclude that M_4 is graphic as in the previous cases. First we have to find a separating cocircuit. Consider row 1 of R_4, that is, the cocircuit $Y = \{e_1, e_8, e_9, e_{10}, e_{11}, e_{12}\}$. The representation matrix for $M_4 \backslash Y$ will be

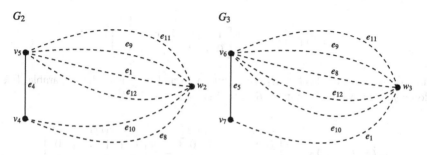

Fig. 5.8 Graphs G_2 and G_3

$$R_4 \setminus Y = \begin{matrix} 1 \\ 6 \\ 7 \end{matrix} \begin{bmatrix} \overset{e_6\ e_7\ e_{13}}{0\ 0\ 0} \\ 1\ 0\ 1 \\ 0\ 1\ 1 \end{bmatrix},$$

and we can see that $R_4 \setminus Y$ cannot be partitioned into at least two blocks, which means that Y is a non-separating cocircuit. Consider row 6, which corresponds to the cocircuit $Y = \{e_6, e_{11}, e_{12}, e_{13}\}$. We get

$$R_4 \setminus Y = \begin{matrix} 1 \\ 6 \\ 7 \end{matrix} \begin{bmatrix} \overset{e_1\ e_7\ e_8\ e_9\ e_{10}}{1\ 0\ 1\ 1\ 1} \\ 0\ 0\ 0\ 0\ 0 \\ 0\ 1\ 0\ 1\ 1 \end{bmatrix},$$

and, as previously, we can see that Y is non-separating. For row 7 of R_4 we have the cocircuit $Y = \{e_7, e_9, e_{10}, e_{11}, e_{12}, e_{13}\}$ and

$$R_4 \setminus Y = \begin{matrix} 1 \\ 6 \\ 7 \end{matrix} \begin{bmatrix} \overset{e_1\ e_8\ e_6}{1\ 1\ 0} \\ 0\ 0\ 1 \\ 0\ 0\ 0 \end{bmatrix},$$

where we see from the partition of the matrix that $M_4 \setminus Y$ has two separators, and namely $B_1 = \{e_1, e_8\}$ and $B_2 = \{e_6\}$. Let $Y_2 = \{e_7, e_9, e_{10}, e_{11}, e_{12}, e_{13}\}$. If the cocircuit of row 7 was also a non-separating cocircuit we would have concluded that M_4 is not a graphic matroid, which would in turn imply that M is not graphic, since graphic matroids are closed under minors as stated in Proposition 4.9. Working as we did before for M, the Y_2-components $M_4.(B_i \cup Y_2)$, for $i = 1, 2$, are given by the following representation matrices:

$$R_4.(B_1 \cup Y_2) = \begin{matrix} 1 \\ 7 \end{matrix} \begin{bmatrix} \overset{e_1\ e_7}{1\ 0} & \overset{e_8\ e_9\ e_{10}\ e_{11}\ e_{12}\ e_{13}}{1\ 1\ 1\ 1\ 1\ 0} \\ 0\ 1 & 0\ 1\ 1\ 1\ 1\ 1 \end{bmatrix}, \tag{5.10}$$

$$R_4.(B_2 \cup Y_2) = \begin{matrix} 6 \\ 7 \end{matrix} \begin{bmatrix} \overset{e_6\ e_7}{1\ 0} & \overset{e_9\ e_{10}\ e_{11}\ e_{12}\ e_{13}}{0\ 0\ 1\ 1\ 1} \\ 0\ 1 & 1\ 1\ 1\ 1\ 1 \end{bmatrix}. \tag{5.11}$$

Initially we have to check if B_1, B_2 partition Y_2 and if Y_2 is bridge-separable. For doing so we have to compute $\pi(B_i, M_4, Y_2)$ for $i = 1, 2$. We will have

$$R_4.(B_1 \cup Y_2)|Y_2 = \begin{matrix} 1 \\ 7 \end{matrix} \begin{bmatrix} \overset{e_7\ e_9\ e_{10}\ e_{11}\ e_{12}\ e_{13}}{0\ 1\ 1\ 1\ 1\ 0} \\ 1\ 1\ 1\ 1\ 1\ 1 \end{bmatrix} \rightarrow \begin{matrix} 1 \\ 7 \end{matrix} \begin{bmatrix} \overset{e_7\ e_9\ e_{10}\ e_{11}\ e_{12}\ e_{13}}{0\ 1\ 1\ 1\ 1\ 0} \\ 1\ 0\ 0\ 0\ 0\ 1 \end{bmatrix},$$

which gives $\pi(B_2, M_4, Y_2) = \{\{e_9, e_{10}, e_{11}, e_{12}\}, \{e_7, e_{13}\}\}$, and

$$R_4.(B_2 \cup Y_2)|Y_2 = \begin{matrix} & \begin{matrix} e_7 & e_9 & e_{10} & e_{11} & e_{12} & e_{13} \end{matrix} \\ \begin{matrix} 6 \\ 7 \end{matrix} & \begin{bmatrix} 0 & 0 & 0 & 1 & 1 & 1 \\ 1 & 1 & 1 & 1 & 1 & 1 \end{bmatrix} \end{matrix} \rightarrow \begin{matrix} & \begin{matrix} e_7 & e_9 & e_{10} & e_{11} & e_{12} & e_{13} \end{matrix} \\ \begin{matrix} 6 \\ 7 \end{matrix} & \begin{bmatrix} 0 & 0 & 0 & 1 & 1 & 1 \\ 1 & 1 & 1 & 0 & 0 & 0 \end{bmatrix} \end{matrix},$$

which give $\pi(B_2, M_4, Y_2) = \{\{e_7, e_9, e_{10}\}, \{e_{11}, e_{12}, e_{13}\}\}$. Therefore, B_1 avoids B_2 and both bridges partition Y_2. Now we have to check whether the binary matroids given by the representation matrices in (5.10) and (5.11) are graphic matroids. Letting $M_5 = M_4.(B_1 \cup Y_2)$, $M_6 = M_4.(B_2 \cup Y_2)$ and their representation matrices $R_5 = R_4.(B_1 \cup Y_2)$ and $R_6 = R_4.(B_2 \cup Y_2)$, we observe that both R_5 and R_6 have at most two nonzero elements in each column, hence, the matroids are graphic. To construct the incidence matrices R_5', R_6' of the graphs G_5, G_6 such that $M(G_5) = M_5$ and $M(G_6) = M_6$, we work as we did previously for constructing R_1', R_2' and R_3'. Specifically, we will have

$$R_5' = \begin{matrix} & \begin{matrix} e_1 & e_8 & e_7 & e_9 & e_{10} & e_{11} & e_{12} & e_{13} \end{matrix} \\ \begin{matrix} v_8 \\ v_9 \\ w_5 \end{matrix} & \begin{bmatrix} 1 & 1 & 0 & 1 & 1 & 1 & 1 & 0 \\ 1 & 1 & 1 & 0 & 0 & 0 & 0 & 1 \\ 0 & 0 & 1 & 1 & 1 & 1 & 1 & 1 \end{bmatrix} \end{matrix}$$

$$R_6' = \begin{matrix} & \begin{matrix} e_6 & e_7 & e_9 & e_{10} & e_{11} & e_{12} & e_{13} \end{matrix} \\ \begin{matrix} v_{10} \\ v_{11} \\ w_6 \end{matrix} & \begin{bmatrix} 1 & 0 & 0 & 0 & 1 & 1 & 1 \\ 1 & 1 & 1 & 1 & 0 & 0 & 0 \\ 0 & 1 & 1 & 1 & 1 & 1 & 1 \end{bmatrix} \end{matrix}.$$

The associated graphs G_5 and G_6 are shown in Fig. 5.9.

Therefore, the matroid M_4 is graphic since it satisfies the conditions of Theorem 5.6. In order to construct the graph G_4 such that $M(G_4) = M_4$, we take the star composition of the matrices R_5' and R_6' at Y_2. Specifically, we will have

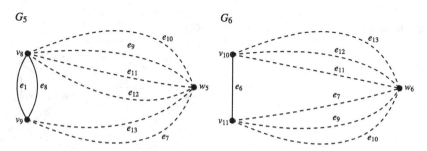

Fig. 5.9 Graphs G_5 and G_6

$$R_5' \circledast_{Y_2} R_6' = \begin{array}{c} \\ v_8 \\ v_9 \\ v_{10} \\ v_{11} \end{array} \begin{array}{c} e_1 \; e_8 \quad e_6 \quad e_7 \; e_9 \; e_{10} \; e_{11} \; e_{12} \; e_{13} \\ \left[\begin{array}{cc|c|cccccc} 1 & 1 & 0 & 0 & 1 & 1 & 1 & 1 & 0 \\ 1 & 1 & 0 & 1 & 0 & 0 & 0 & 0 & 1 \\ \hline 0 & 0 & 1 & 0 & 0 & 0 & 1 & 1 & 1 \\ 0 & 0 & 1 & 1 & 1 & 1 & 0 & 0 & 0 \end{array} \right] \end{array}.$$

Although $R_5' \circledast_{Y_2} R_6'$ is the incidence matrix of a graph corresponding to the graphic matroid M_4, among all possible graphical representations of M_4, we want the one where Y_1 is the star of a vertex, so as to be able to compose its graph with the one derived from the respective star composition of G_1, G_2 and G_3. In this case we see that in $R_5' \circledast_{Y_2} R_6'$ the cocircuit Y_1 appears in row v_8, therefore, we have the following incident matrix of G_4 after permuting some columns and rows

$$R_4' = \begin{array}{c} \\ v_9 \\ v_{10} \\ v_{11} \\ v_8 \end{array} \begin{array}{c} e_6 \; e_7 \; e_{13} \quad e_1 \; e_8 \; e_9 \; e_{10} \; e_{11} \; e_{12} \\ \left[\begin{array}{ccc|cccccc} 0 & 1 & 1 & 1 & 1 & 0 & 0 & 0 & 0 \\ 1 & 0 & 1 & 0 & 0 & 0 & 0 & 1 & 1 \\ 1 & 1 & 0 & 0 & 0 & 1 & 1 & 0 & 0 \\ \hline 0 & 0 & 0 & 1 & 1 & 1 & 1 & 1 & 1 \end{array} \right] \end{array}.$$

The graph G_4 is given in Fig. 5.10.

We can conclude that the Y_1-components M_1, M_2, M_3, and M_4 are all graphic matroids, hence, the matroid M is graphic, since it fulfills the conditions of Theorem 5.6. In order to construct the graph G which represents M we have to construct two graphs as the star compositions of the graphs that correspond to the two collections of avoiding bridges, where Y_1 appears as a star of a vertex in both graphs, and then take the star composition of these. This has been done already for G_4, so it remains to take the star compositions of the graphs G_1, G_2, and G_3. The star compositions will have to be done in pairs, since for each incident matrix which results from a star composition we will probably have to apply elementary row operations to compute an equivalent incident matrix, where Y_1 appears as a star of a vertex. For G_1 and G_2, we have the incidence matrix

Fig. 5.10 Graph G_4 G_4

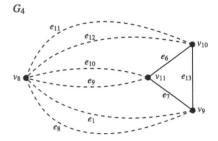

$$R'_1 \circledast_{Y_1} R'_2 = \begin{array}{c} \\ v_1 \\ v_2 \\ v_3 \\ v_4 \\ v_5 \end{array} \begin{array}{c} \begin{matrix} e_2 & e_3 & e_{14} & & e_4 & & e_1 & e_8 & e_9 & e_{10} & e_{11} & e_{12} \end{matrix} \\ \left[\begin{array}{ccc|c|cccccc} 1 & 0 & 1 & 0 & 0 & 1 & 0 & 1 & 0 & 0 \\ 0 & 1 & 1 & 0 & 0 & 0 & 1 & 0 & 1 & 0 \\ 1 & 1 & 0 & 0 & 1 & 0 & 0 & 0 & 0 & 1 \\ 0 & 0 & 0 & 1 & 0 & 1 & 0 & 1 & 0 & 0 \\ 0 & 0 & 0 & 1 & 1 & 0 & 1 & 0 & 1 & 1 \end{array} \right] \end{array}$$

where we see that Y_1 does not appear as a cocircuit. Adding row v_4 to rows v_5 and v_1 we get the incident matrix for the graph $G_{1,2}$

$$R'_{1,2} = \begin{array}{c} \\ v_1 \\ v_2 \\ v_3 \\ v_4 \\ v_5 \end{array} \begin{array}{c} \begin{matrix} e_2 & e_3 & e_{14} & e_4 & & e_1 & e_8 & e_9 & e_{10} & e_{11} & e_{12} \end{matrix} \\ \left[\begin{array}{cccc|cccccc} 1 & 0 & 1 & 1 & 0 & 0 & 0 & 0 & 0 & 0 \\ 0 & 1 & 1 & 0 & 0 & 0 & 1 & 0 & 1 & 0 \\ 1 & 1 & 0 & 0 & 1 & 0 & 0 & 0 & 0 & 1 \\ 0 & 0 & 0 & 1 & 0 & 1 & 0 & 1 & 0 & 0 \\ 0 & 0 & 0 & 0 & 1 & 1 & 1 & 1 & 1 & 1 \end{array} \right] \end{array}$$

where Y_1 is a star of the vertex v_5. The graph $G_{1,2}$ is given in Fig. 5.11. Note that the graph $G_{1,2}$ can be obtained from the graph that corresponds to the incidence matrix $R'_1 \circledast_{Y_1} R'_2$ by twisting about the vertices v_1 and v_5. Continuing in this way we can compose $R'_{1,2}$ with R'_3 at Y_1 to get

$$R'_{1,2} \circledast_{Y_1} R'_3 = \begin{array}{c} \\ v_1 \\ v_2 \\ v_3 \\ v_4 \\ v_6 \\ v_7 \end{array} \begin{array}{c} \begin{matrix} e_2 & e_3 & e_{14} & e_4 & & e_5 & & e_1 & e_8 & e_9 & e_{10} & e_{11} & e_{12} \end{matrix} \\ \left[\begin{array}{cccc|c|cccccc} 1 & 0 & 1 & 1 & 0 & 0 & 0 & 0 & 0 & 0 & 0 \\ 0 & 1 & 1 & 0 & 0 & 0 & 0 & 1 & 0 & 1 & 0 \\ 1 & 1 & 0 & 0 & 0 & 1 & 0 & 0 & 0 & 0 & 1 \\ 0 & 0 & 0 & 1 & 0 & 0 & 1 & 0 & 1 & 0 & 0 \\ 0 & 0 & 0 & 0 & 1 & 0 & 1 & 1 & 1 & 1 & 1 \\ 0 & 0 & 0 & 0 & 1 & 1 & 0 & 0 & 0 & 0 & 0 \end{array} \right] \end{array} .$$

Adding row v_7 to rows v_6 and v_3, and interchanging rows v_6 and v_7 we get the incidence matrix of $G_{1,2,3}$ where Y_1 appears as a cocircuit

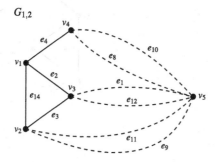

Fig. 5.11 Graph $G_{1,2}$

$$R'_{1,2,3} = \begin{array}{c} \\ v_1 \\ v_2 \\ v_3 \\ v_4 \\ v_7 \\ v_6 \end{array} \begin{array}{c} \begin{array}{ccccc} e_2 & e_3 & e_{14} & e_4 & e_5 \end{array} \begin{array}{cccccc} e_1 & e_8 & e_9 & e_{10} & e_{11} & e_{12} \end{array} \\ \left[\begin{array}{ccccc|cccccc} 1 & 0 & 1 & 1 & 0 & 0 & 0 & 0 & 0 & 0 & 0 \\ 0 & 1 & 1 & 0 & 0 & 0 & 0 & 1 & 0 & 1 & 0 \\ 1 & 1 & 0 & 0 & 1 & 0 & 0 & 0 & 0 & 0 & 1 \\ 0 & 0 & 0 & 1 & 0 & 0 & 1 & 0 & 1 & 0 & 0 \\ 0 & 0 & 0 & 0 & 1 & 1 & 0 & 0 & 0 & 0 & 0 \\ 0 & 0 & 0 & 0 & 0 & 1 & 1 & 1 & 1 & 1 & 1 \end{array} \right] \end{array},$$

while the graph $G_{1,2,3}$ is given in Fig. 5.12.

Finally, we compose $R'_{1,2,3}$ with R'_4 at Y_1 to get the incidence matrix of the graph G shown in Fig. 5.13.

$$R'_{1,2,3} \circledast_{Y_1} R'_4 = \begin{array}{c} \\ v_1 \\ v_2 \\ v_3 \\ v_4 \\ v_7 \\ v_9 \\ v_{10} \\ v_{11} \end{array} \begin{array}{c} \begin{array}{ccccc} e_2 & e_3 & e_{14} & e_4 & e_5 \end{array} \begin{array}{ccc} e_6 & e_7 & e_{13} \end{array} \begin{array}{cccccc} e_1 & e_8 & e_9 & e_{10} & e_{11} & e_{12} \end{array} \\ \left[\begin{array}{ccccc|ccc|cccccc} 1 & 0 & 1 & 1 & 0 & 0 & 0 & 0 & 0 & 0 & 0 & 0 & 0 & 0 \\ 0 & 1 & 1 & 0 & 0 & 0 & 0 & 0 & 0 & 0 & 1 & 0 & 1 & 0 \\ 1 & 1 & 0 & 0 & 1 & 0 & 0 & 0 & 0 & 0 & 0 & 0 & 0 & 1 \\ 0 & 0 & 0 & 1 & 0 & 0 & 0 & 0 & 0 & 1 & 0 & 1 & 0 & 0 \\ 0 & 0 & 0 & 0 & 1 & 0 & 0 & 0 & 1 & 0 & 0 & 0 & 0 & 0 \\ 0 & 0 & 0 & 0 & 0 & 0 & 1 & 1 & 1 & 1 & 0 & 0 & 0 & 0 \\ 0 & 0 & 0 & 0 & 0 & 1 & 0 & 1 & 0 & 0 & 0 & 0 & 1 & 1 \\ 0 & 0 & 0 & 0 & 0 & 1 & 1 & 0 & 0 & 0 & 1 & 1 & 0 & 0 \end{array} \right] \end{array}.$$

Note that Y_1 is a bond of G, however, it does not appear as the star of a vertex in the incidence matrix, and we cannot find a set of elementary row operations to make this so. This is because we have overlapping bridges of Y_1 in the composing graphs $G_{1,2,3}$ and G_4, therefore, Theorem 5.5 is no longer applicable. The reader should be able to verify that the columns $e_8, \ldots, e_{12}$ of the matrix R in (5.5) are characteristic vectors of fundamental cycles with respect to the spanning tree of G induced by $\{e_1, \ldots, e_7\}$.

Fig. 5.12 Graph $G_{1,2,3}$

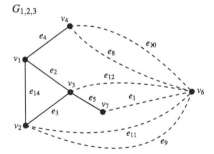

$G_{1,2,3}$

Fig. 5.13 Graph G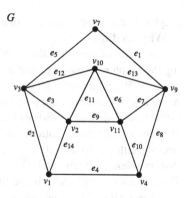

5.5 Notes

With the exception of Theorem 5.2 and its corollary, all the results that appear in Sects. 5.1 and 5.2 of this chapter are more or less dual versions of the results from Tutte (1959, 1960, 1965). Tutte preferred to work with cographic matroids, that is, by a graphic matroid he was referring to the matroid which is isomorphic to the bond matroid of a graph. For instance in (Tutte 1959), the bridges are defined as the elementary separators of M/Y for some circuit Y of a *cographic* matroid M. For the most part of this chapter we chose to adopt Tutte's original terminology for the convenience of the reader that wishes to consult the original works. For anything else in matroids and graphs, we tried to use standard terminology and notation, as the one used by Oxley (1992) and Diestel (2006). For example, Tutte called the circuits of a matroid *points* in (Tutte 1959) or *atoms* in (Tutte 1965), and the elements of the ground set *cells*, while the cycles of a graph *circuits* in (Tutte 1959) or *polygons* in (Tutte 1965). The majority of the proofs given in this chapter are based on the original proofs by Tutte (1959), with various simplifications and clarifications, or alternative parts in the proof. In some results a different proof is given altogether.

Apart from the Theorem 4.12 on excluded minors and the decomposition result in Theorem 5.6 by Tutte, there are also other characterizations for graphic matroids in the literature. Fournier (1974) provided a characterization based on a condition for the intersection of any three cocircuits. Mighton (2008) proved that this condition combined with bridge-separability of fundamental cocircuits in any given base are necessary and sufficient conditions for a binary matroid to be graphic. This leads also to an efficient algorithm for checking whether a binary matroid is graphic. Recently, Geelen and Gerards (2013) provided a characterization based on the existence of a solution of a linear system formed by the binary representation matrix.

In the proof of Theorem 5.5 we used the fact that if a graph is obtained by another graph by a sequence of twistings, then both graphs have the same cycle matroid. The operation of twisting was introduced in Whitney (1933), where the reverse direction of the previous statement is proved, that is, all graph representations of a graphic matroid are related by twistings. Twistings are also known in the literature as **Whitney-flips**.

A number of polynomial-time algorithms have appeared in the literature for solving the graph realization problem, and can be found in the works of Auslander and Trent (1959), Cunningham (1982), Tamari 1977), and Rajappan and Stone (1971) to name a few. Currently, the best known algorithms are those by Bixby and Wagner (1988) and Fujishige (1980) which are almost linear-time in the number of nonzero elements of the representation matrix. One of the main applications of a graph realization algorithm is to identify and convert linear programming problems into network flow problems, which can be solved efficiently. This has been done by Bixby and Cunningham (1980), which use the results by Tutte (1959). An algorithm to test whether a general matroid, not necessarily binary, given by an independence oracle is graphic is given by Seymour (1981).

Chapter 6
Signed-Graphic Matroids

Signed graphs are ordinary graphs with an additional structure that results from the assignment of signs to the edges of the graph. In this chapter we will demonstrate how signed-graphic matroids arise from signed graphs and present a decomposition theory which extends the one presented in Chap. 5 for graphic matroids.

6.1 Signed Graphs

Let us extend the definition of a graph given in Sect. 2.1, by allowing the edges to be n-tuples of the vertex set for $n = 0, 1, 2$. Therefore, given two distinct vertices $v, u \in V$ an edge e can be one of the following four different types: a **link** $e = (v, u)$, a loop $e = (v, v)$, a **half-edge** $e = (v)$, and a **loose-edge** $e = \emptyset$. Given an **underlying graph** $G(V, E)$, a **signed graph** is defined as $\Sigma = (G, \sigma)$, where σ is a sign function $\sigma : E(G) \rightarrow \{\pm 1\}$, such that $\sigma(e) = -1$ if e is a half-edge and $\sigma(e) = +1$ if e is a loose-edge. Therefore, a signed graph is a graph where the edges are labelled as positive or negative while all the half-edges are negative and all the loose-edges are positive. We denote by $V(\Sigma)$ and $E(\Sigma)$ the vertex and edge sets of a signed graph Σ, respectively. All operations on signed graphs may be defined through the corresponding operation on the underlying graph and the sign function. In the following definitions assume that we have a signed graph $\Sigma = (G, \sigma)$. The operation of **switching** at a vertex v results in a new signed graph $(G, \bar{\sigma})$ where $\bar{\sigma}(e) = -\sigma(e)$ for each link e incident to v, while $\bar{\sigma}(e) = \sigma(e)$ for all other edges. We say that $\bar{\sigma}$ is a switching at vertex v. **Deletion** of an edge e is defined as $\Sigma \backslash \{e\} = (G \backslash \{e\}, \sigma)$. The **contraction** of an edge e consists of three cases:

(i) If e is a positive loop, then $\Sigma / \{e\} = (G \backslash \{e\}, \sigma)$.
(ii) If e is a half-edge, negative loop or a positive link, then $\Sigma / \{e\} = (G / \{e\}, \sigma)$.
(iii) If e is a negative link, then $\Sigma / \{e\} = (G / \{e\}, \bar{\sigma})$ where $\bar{\sigma}$ is a switching at either one of the end-vertices of e.

L. S. Pitsoulis, *Topics in Matroid Theory*,
SpringerBriefs in Optimization, DOI: 10.1007/978-1-4614-8957-3_6,
© Leonidas S. Pitsoulis 2014

The connectivity of a signed graph is the connectivity of its underlying graph. So a signed graph is k-connected if and only if its underlying graph is k-connected. The sign of a cycle in a signed graph is the product of the signs of its edges, so we have a **positive cycle** if the number of negative edges in the cycle is even, otherwise the cycle is a **negative cycle**. Both negative loops and half-edges are negative cycles with a single edge. A signed graph is called **balanced** if it contains no negative cycles. A vertex $v \in V(\Sigma)$ is called a **balancing vertex** if $\Sigma \backslash \{v\}$ is balanced.

The **incidence** matrix of a signed graph $\Sigma(G, \sigma)$, is a $|V(G)| \times |E(G)|$ matrix $A_\Sigma \in GF(3)$ with columns $\mathbf{a}_e$ for each $e \in E(\Sigma)$ defined as follows:

(i) if $e = (v, w)$ is a positive link then $\mathbf{a}_e$ is $\begin{array}{c} e \\ v \begin{bmatrix} 1 \\ w \end{bmatrix} \begin{array}{c} 2 \end{array}$ or $\begin{array}{c} e \\ v \begin{bmatrix} 2 \\ w \end{bmatrix} \begin{array}{c} 1 \end{array}$,

(ii) if $e = (v, w)$ is a negative link then $\mathbf{a}_e$ is $\begin{array}{c} e \\ v \begin{bmatrix} 1 \\ w \end{bmatrix} \begin{array}{c} 1 \end{array}$ or $\begin{array}{c} e \\ v \begin{bmatrix} 2 \\ w \end{bmatrix} \begin{array}{c} 2 \end{array}$,

(iii) if $e = (v)$ is a half-edge or $e = (v, v)$ is a negative loop then $\mathbf{a}_e$ is $\begin{array}{c} e \\ v \begin{bmatrix} 1 \end{bmatrix} \end{array}$ or

$\begin{array}{c} e \\ v \begin{bmatrix} 2 \end{bmatrix} \end{array}$,

(iv) if $e = (v, v)$ is a positive loop then $\mathbf{a}_e = 0$,

where all the other unspecified elements in the columns are zero. So negative edges have columns with two 1's or two 2's, while positive edges have a 1 and a 2 in the rows corresponding to the end-vertices of the edge. Half-edges and negative loops have columns that have a 1 or a 2, and positive loops are represented by zero columns. The reason behind the alternative columns in the definition of the incidence matrix is that linear independence is not affected under column scaling in $GF(3)$ and in each case the alternative column is produced after multiplying by 2. A signed graph Σ and its incidence matrix A_Σ are illustrated in Fig. 6.1, where negative edges are shown in light gray color, and positive edges by solid black color.

A **bidirected** graph $\overrightarrow{\Sigma}$ is a signed graph $\Sigma(G, \sigma)$ with an orientation o applied to the underlying graph G such that $-o(e, v)o(e, w) = \sigma(e)$ for any edge $e = (v, w) \in E(\Sigma)$. Therefore, positive edges will have end-vertices with different signs and negative edges with the same sign. The sign of an end-vertex of a half-edge is always positive. Bidirected graphs can be thought of as the oriented signed-graphic analog of directed graphs with edges that can be directed and bidirected. The **incidence** matrix of a bidirected graph $\overrightarrow{\Sigma}$, is a $|V(G)| \times |E(G)|$ matrix $A_{\overrightarrow{\Sigma}} = (a_{ij}) \in \mathbb{R}$ defined as follows:

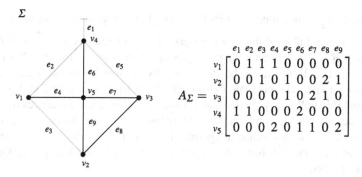

Fig. 6.1 A signed graph and its incidence matrix

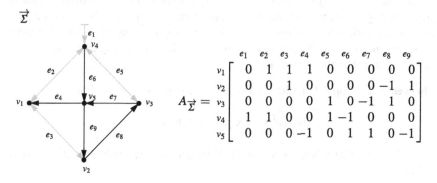

Fig. 6.2 A bidirected graph and its incidence matrix

$$a_{ve} = \begin{cases} +1 & \text{if vertex } v \text{ is the head of the non-loop arc } e, \\ -1 & \text{if vertex } v \text{ is the tail of the non-loop arc } e, \\ +2 & \text{if vertex } v \text{ is the head of the loop arc } e, \\ -2 & \text{if vertex } v \text{ is the tail of the loop arc } e, \\ 0 & \text{otherwise.} \end{cases}$$

Note that in the case of a vertex v being both the tail and the head of a loop arc e, then by the above definition $a_{ve} = 0$. The bidirected graph and its associated incidence matrix so obtained for an orientation of the signed graph in Fig. 6.1 is given in Fig. 6.2. Applying elementary row operations to $A_{\vec{\Sigma}}$ we get

$$
\begin{array}{c}
\quad\;\; e_1\, e_2\, e_3\, e_4\, e_5 \quad e_6 \quad e_7 \quad e_8 \quad e_9 \\
\begin{array}{c} e_1 \\ e_2 \\ e_3 \\ e_4 \\ e_5 \end{array}
\left[\begin{array}{ccccc|cccc}
1 & 0 & 0 & 0 & 0 & -2 & 0 & -2 & 2 \\
0 & 1 & 0 & 0 & 0 & 1 & 1 & 1 & -2 \\
0 & 0 & 1 & 0 & 0 & 0 & 0 & -1 & 1 \\
0 & 0 & 0 & 1 & 0 & -1 & -1 & 0 & 1 \\
0 & 0 & 0 & 0 & 1 & 0 & -1 & 1 & 0
\end{array}\right] = [\, I \mid B_{\vec{\Sigma},T} \,].
\end{array}
\tag{6.1}
$$

The matrix $B_{\overrightarrow{\Sigma},T}$ is called the **binet** matrix of $\overrightarrow{\Sigma}$. Specifically, we have that if $[\ R\ |\ S\ ]$ is a full row incidence matrix of a bidirected graph $\overrightarrow{\Sigma}$ and R is a basis, then $R^{-1}S$ is a binet matrix. In general an $m \times n$ binet matrix has entries in $\{0, \pm\frac{1}{2}, \pm1, \pm2\}$, but it can be shown that by at most $2m$ pivots it can be converted into an integral binet matrix (Kotnyek 2002). Moreover, note that $A_{\overrightarrow{\Sigma}} = A_\Sigma$ mod 3 and this will always be true, as for the case of incidence matrices of graphs and directed graphs. A combinatorial algorithm to compute the entries of a binet matrix associated with a bidirected graph, similar to the one described for network matrices in Example 4.1, is given independently by Appa and Kotnyek (2006) and Zaslavsky (2006).

Binet matrices are important in optimization since they define a well-solved class of combinatorial optimization problems. In Appa et al. (2007) it is proved that linear programming problems with binet constraint matrices can be solved via the generalized network simplex algorithm and integer programming problems with binet constraint matrices can be converted into a generalized matching problem. Specifically, we have the following result.

Theorem 6.1 (Appa et al. (2007)) *The integer programming problem*

$$\max\{\mathbf{c}^T\mathbf{x} : B\mathbf{x} \leq \mathbf{b}, \mathbf{x} \geq \mathbf{0}, \mathbf{x} \in \mathbb{Z}\}$$

where B is an integral binet matrix and $\mathbf{b}$ an integral vector, can be solved in strongly polynomial time.

Moreover, in (Appa et al. 2007) it is also proved that integral binet matrices have strong Chvátal rank 1. Therefore, a polynomial time recognition algorithm for binet matrices would provide the means of recognizing the aforementioned classes of well-solved integer programming problems.

6.2 Signed-Graphic Matroids

Given a signed graph $\Sigma(G, \sigma)$ we wish to define a matroid that captures its structure much in the same way as we did for graphic matroids. The obvious cycle matroid of the underlying graph $M(G)$ is of little importance since it does not take into account the sign function σ. In Theorems 4.1 and 4.2 we saw that the cycle matroid of a graph is the vector matroid of the associated incidence matrix. The situation is similar for signed graphs. Applying elementary row operations in $GF(3)$ and column interchanges to A_Σ in Fig. 6.1, we can obtain the following standard representation matrix of $M[A_\Sigma]$:

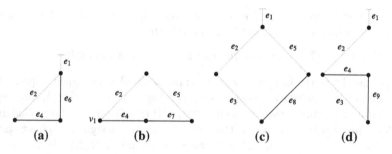

Fig. 6.3 Fundamental circuits in $M[A_\Sigma]$. (a) $C(e6, T)$ (b) $C(e7, T)$ (c) $C(e8, T)$ (d) $C(e9, T)$

$$
\begin{array}{c}
\begin{array}{ccccccccc} e_1 & e_2 & e_3 & e_4 & e_5 & e_6 & e_7 & e_8 & e_9 \end{array} \\
\begin{array}{c} e_1 \\ e_2 \\ e_3 \\ e_4 \\ e_5 \end{array}
\left[\begin{array}{ccccc|cccc}
1 & 0 & 0 & 0 & 0 & 1 & 0 & 1 & 2 \\
0 & 1 & 0 & 0 & 0 & 1 & 1 & 1 & 1 \\
0 & 0 & 1 & 0 & 0 & 0 & 0 & 2 & 1 \\
0 & 0 & 0 & 1 & 0 & 2 & 2 & 0 & 1 \\
0 & 0 & 0 & 0 & 1 & 0 & 2 & 1 & 0
\end{array}\right] = [\, I \mid B_{\Sigma,T} \,]. \qquad (6.2)
\end{array}
$$

The matrix $B_{\Sigma,T}$ is a compact representation matrix of $M[A_\Sigma]$. The elements of $T = \{e_1, e_2, e_3, e_4, e_5\}$ constitute a base for $M[A_\Sigma]$, while the characteristic vectors of the other columns represent fundamental circuits with respect to the base T. In Fig. 6.3 we can see the subgraphs of Σ induced by these fundamental circuits of $M[A_\Sigma]$. We can distinguish three types of graphs that correspond to circuits in $M[A_\Sigma]$; a positive cycle as in the case of $C(e_7, T)$, two negative cycles with a common vertex as in the cases of $C(e_6, T)$, $C(e_8, T)$, or two vertex-disjoint negative cycles joined by a path as in the case of $C(e_9, T)$. The following theorem states that edge sets in a signed graph that induce the aforementioned types of subgraphs constitute a family of circuits of a matroid which we will call **signed-graphic matroid**.

Theorem 6.2 (Zaslavsky 1982) *Given a signed graph Σ let $\mathscr{C}$ be the family of edge sets inducing a subgraph in Σ which is either:*

(i) *a positive cycle, or*
(ii) *two negative cycles which have exactly one common vertex, or*

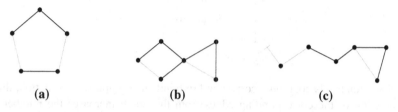

Fig. 6.4 Circuits in a signed graph Σ (a) Positive cycle (b) Tight handcuff (c) Loose handcuff

(iii) *two vertex-disjoint negative cycles connected by a path which has no common vertex with the cycles apart from its end-vertices.*

Then $M(\Sigma) = (E(\Sigma), \mathscr{C})$ is a matroid on $E(\Sigma)$ with circuit family $\mathscr{C}$.

$M(\Sigma)$ is also known in the literature as the **frame matroid** of the signed graph Σ. The subgraphs of Σ induced by the edges corresponding to a circuit of $M(\Sigma)$ are called the **circuits** of Σ, while those described by (ii) and (iii) of Theorem 6.2 are called **tight** and **loose handcuffs** respectively. The types of circuits in a signed graph are shown in Fig. 6.4. Having established a link between the circuits of $M(\Sigma)$ with the circuits of Σ we can apply similar techniques as the ones used in the proofs of Theorems 4.1 and 4.2, to establish representability of signed-graphic matroids over $GF(3)$ and $\mathbb{R}$. The signs of the edges in the circuits of the graph will determine the coefficients of a linear combination of linearly dependent columns in the incidence matrix. We have the following theorem regarding the representability of signed-graphic matroids:

Theorem 6.3 *(Papalamprou and Pitsoulis 2012) Given a signed graph Σ*

(i) *A_Σ is a representation of $M(\Sigma)$ in $GF(3)$, and*
(ii) *$A_{\overrightarrow{\Sigma}}$ is a representation of $M(\Sigma)$ in $\mathbb{R}$, where $\overrightarrow{\Sigma}$ is a bidirected graph obtained by an arbitrary orientation of Σ.*

A non-constructive proof of the fact that signed-graphic matroids are ternary is given in (Zaslavsky 1982). Observe that for the matrices in (6.1) and (6.2) we have that $B_{\Sigma,T} = B_{\overrightarrow{\Sigma},T}$ mod 3. As the following theorem states this defines the relationship between ternary and real representations of a signed-graphic matroid.

Theorem 6.4 *(Papalamprou and Pitsoulis 2012) Let B be an integral binet matrix and $M(\Sigma)$ be the signed-graphic matroid represented by B over $\mathbb{R}$. The matrix $B' = B$ mod 3 is a compact representation matrix of $M(\Sigma)$ over $GF(3)$.*

Let us turn our attention now to the cocircuits of signed-graphic matroids. We know by Theorem 4.9 that the rows of the standard representation matrix (6.2) are characteristic vectors of the fundamental cocircuits of $M[A_\Sigma]$ with respect to the base $E(\Sigma) - T$. We will have the following cocircuits:

$$C_1^* = \{e_1, e_6, e_8, e_9\},$$
$$C_2^* = \{e_2, e_6, e_7, e_8, e_9\},$$
$$C_3^* = \{e_3, e_8, e_9\},$$
$$C_4^* = \{e_4, e_6, e_7, e_9\},$$
$$C_5^* = \{e_5, e_7, e_8\}.$$

In graphic matroids cocircuits correspond to bonds in a graph representation, thus, the deletion of the corresponding edges from the graph increases the number of connected components and these sets of edges are minimal with respect to this property. In Fig. 6.5 we can see the subgraphs of Σ so obtained by the deletion of

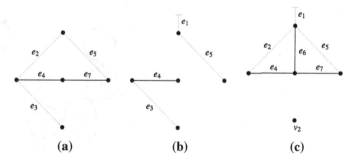

Fig. 6.5 Fundamental cocircuits in $M[A_\Sigma]$. (a) $\Sigma \backslash C_1^*$ (b) $\Sigma \backslash C_2^*$ (c) $\Sigma \backslash C_3^*$

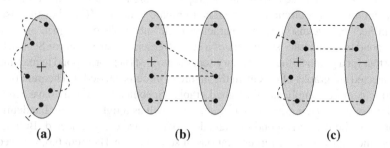

Fig. 6.6 Bonds in a signed graph Σ. (a) Balancing bond (b) Unbalancing bond (c) Double bond

the cocircuits C_1^*, C_2^* and C_3^*, while C_4^* and C_5^* are similar to C_3^* as they are also stars of a vertex. We observe that in all cases the cocircuits consist of minimal sets of edges whose deletion increases the number of balanced components. Moreover, the resulting graph can be connected as for the case of C_1^* or disconnected as in the other cases. The graph $\Sigma \backslash C_1^*$ has one connected balanced component, the graph $\Sigma \backslash C_2^*$ has the balanced component induced by the edges e_3 and e_4, and the graph $\Sigma \backslash C_3^*$ has the trivial graph v_2 as the balanced component created by the deletion of C_3^* which is the star of this vertex. The following theorem characterizes the sets of edges in a signed graph Σ that correspond to cocircuits of $M(\Sigma)$.

Theorem 6.5 (Zaslavsky 1982) *Given a signed graph Σ and its corresponding matroid $M(\Sigma)$, $Y \subseteq E(\Sigma)$ is a cocircuit of $M(\Sigma)$ if and only if Y is a minimal set of edges whose deletion increases the number of balanced components of Σ.*

The sets of edges defined in Theorem 6.5 are called the **bonds** of a signed graph and are illustrated in Fig. 6.6, where the edges of the bond are shown with dashed lines and continuous shaded areas correspond to connected components. As we have seen in Fig. 6.5 we have three types of bonds in a signed graph. A **balancing** bond is a minimal set of edges $Y \subseteq E(\Sigma)$ such that $\Sigma \backslash Y$ is balanced and connected. If $\Sigma \backslash Y$ has a balanced component and an unbalanced component then it is a **double** bond, while if each edge of Y in a double bond is neither a loop nor a half-edge and has its end-vertices contained in both components, then it is called an **unbalancing** bond.

Fig. 6.7 B-necklace

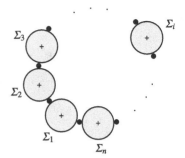

So far we have characterized the subgraphs in a signed graph Σ that correspond to the circuits and cocircuits of the signed-graphic matroid $M(\Sigma)$. For the discussion in Sect. 6.3 we also need to characterize the subgraphs of a signed graph that correspond to elementary separators of a signed-graphic matroid. From Sect. 5.1 we know that for a graph G the elementary separators of $M(G)$ correspond to maximally 2-connected subgraphs of G. Given the nature of circuits defined in Theorem 6.2 the situation is more complicated in signed graphs. Specifically, we may have a signed graph whose underlying graph is 1-connected, such as a tight or loose handcuff for instance, while the corresponding signed-graphic matroid is connected. Before we present the signed-graphic representations of separators in Theorem 6.6, we need to provide some necessary definitions. Given a signed graph $\Sigma(G, \sigma)$ define a **block** as a maximally 2-connected subgraph of G. Any block which is unbalanced or lies on a path between two unbalanced blocks is called **inner** block, while any other block is called **outer**. The **core** of Σ is the union of all inner blocks. A **B-necklace** is a special type of 2-connected unbalanced signed graph, which is composed of balanced blocks Σ_i joined in a cyclic fashion as illustrated in Fig. 6.7. So, while each signed subgraph Σ_i of the B-necklace is balanced, there exists a negative cycle that passes through all the subgraphs which makes their union unbalanced. The unique common vertex between consecutive subgraphs in a B-necklace is called **vertex of attachment**. In the following theorem the subgraphs in a signed graph that correspond to elementary separators of the signed-graphic matroid are characterized.

Theorem 6.6 (Zaslavsky 1991) *Let Σ be a connected signed graph. The elementary separators of $M(\Sigma)$ are the edge sets of each outer block and the core, except when the core is a B-necklace where each block in the B-necklace is also an elementary separator.*

The subgraphs defined in Theorem 6.6 will be called the **separates** of a signed graph. Given the Definition 4.3 of separators in a matroid and the nature of circuits in signed graphs, we can see why the separates in a signed graph are the subgraphs described in Theorem 6.6. The circuits of a signed graph are partitioned by the outer blocks and the core, where each outer block contains only positive cycles, while the core contains positive cycles and handcuffs. This is a key feature of connected signed graphs where the part of the graph which contains all the negative cycles appears as one connected component in the corresponding signed-graphic matroid, since any two negative cycles form a circuit.

Let us state some fundamental properties of signed-graphic matroids with respect to certain graph operations. By Theorems 6.3 and 4.10 we can prove the following signed-graphic counterpart of Theorem 4.9 which states that signed-graphic matroids is a minor-closed class.

Proposition 6.1 *If Σ is a signed graph then*

$$M(\Sigma)\backslash X/Y = M(\Sigma\backslash X/Y)$$

for all $X, Y \subseteq E(\Sigma)$.

The proofs of the following two propositions can be derived from the results in (Slilaty and Qin 2007; Zaslavsky 1982, 1991). Proposition 6.2 states operations on a signed graph that do not alter its family of circuits.

Proposition 6.2 *Let Σ be a signed graph. If Σ'*

(i) *is obtained from Σ by switchings, or*
(ii) *is the twisted graph of Σ about (u, v) with Σ_1, Σ_2 the twisting parts of Σ, where Σ_1 (or Σ_2) is balanced or all of its negative cycles contain u and v,*

   then $M(\Sigma) = M(\Sigma')$.

Note the additional conditions in (ii) of Proposition 6.2, in contrast with graphs where the unconditional application of twistings did not change the family of cycles of the graph. The following proposition states some relatively simple necessary conditions upon which a signed graph has a graphic matroid.

Proposition 6.3 *Let $\Sigma(G, \sigma)$ be a signed graph. If Σ*

(i) *is balanced then $M(\Sigma) = M(G)$,*
(ii) *has no negative cycles other than negative loops and half-edges then $M(\Sigma) = M(G)$,*
(iii) *has a balancing vertex, then $M(\Sigma) = M(G')$ where G' is obtained from G by adding a new vertex v and replacing all negative loops and half-edges by links that connect to v.*

Note that by (iii) of Proposition 6.3 a B-necklace has a graphic matroid since each vertex of attachment is a balancing vertex.

We conclude this introductory section on signed graphs by stating the following problem as a generalization of Problem 5.1 for graphs.

Problem 6.1 (Signed Graph Realization) Given a matrix R in $GF(3)$ find a signed graph Σ such that $M[R] = M(\Sigma)$, or decide that no such signed graph exists.

An algorithm for solving Problem 6.1 together with an appropriate signing scheme would imply a recognition algorithm for binet matrices.

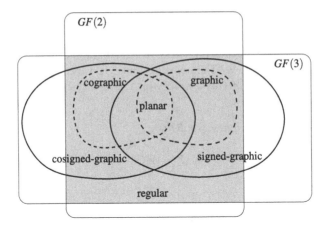

Fig. 6.8 Graphic and signed-graphic matroid representability

6.3 Binary Signed-Graphic Matroids

By Theorem 6.3 we know that signed-graphic matroids are ternary. Moreover, since any graph G is a signed graph $\Sigma(G, \sigma)$ with $\sigma(e) = +1$ for all $e \in E(G)$, we have that any graphic matroid is a signed-graphic matroid. Hence, signed-graphic matroids can be binary. Zaslavsky (1982) has shown that signed-graphic matroids are representable over all fields with characteristic not 2. Combining the above with the results of Pagano (1998) we have the following with respect to the representability of a signed-graphic matroid M:

(i) M is representable over all fields of characteristic not 2;
(ii) if M is representable over $GF(2)$, then it is representable over all fields;
(iii) if M is representable over $GF(4)$ but not $GF(2)$, then it is representable over all fields except $GF(2)$.

In Fig. 6.8 the relationship between the classes of graphic and signed-graphic matroids with respect to representability in the binary and ternary fields is depicted. Any graphic matroid is signed-graphic, and any cographic matroid is cosigned-graphic. There exist binary signed-graphic matroids which are not graphic, such as the matroids $M^*(K_{3,3})$ and $M^*(K_5)$ whose signed graph representations are depicted in Fig. 6.9. Another well-known binary signed-graphic matroid which is neither graphic nor cographic, is the matroid R_{10} with a binary compact representation matrix given in (6.3)

$$\begin{bmatrix} 1 & 1 & 1 & 1 & 1 \\ 1 & 1 & 1 & 0 & 0 \\ 1 & 0 & 1 & 1 & 0 \\ 1 & 0 & 0 & 1 & 1 \\ 1 & 1 & 0 & 0 & 1 \end{bmatrix}, \tag{6.3}$$

Fig. 6.9 Signed graph representations of the matroids $M^*(K_{3,3})$ and $M^*(K_5)$. (a) $\Sigma_{3,3}$ (b) Σ_5

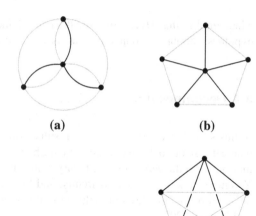

(a)

(b)

Fig. 6.10 Signed graph representation of the matroid R_{10}

and a signed graph representation depicted in Fig. 6.10. By a well-known theorem of Seymour (1980) we know that any regular matroid admits a decomposition into graphic and cographic matroids and copies of R_{10}. A connected signed graph is called **tangled** if it has no balancing vertex and no two vertex-disjoint negative cycles. Therefore, the only possible types of circuits in a tangled signed graph are positive cycles and tight handcuffs. Alternatively, the only possible bonds in tangled signed graphs are balancing bonds and unbalancing bonds. The next theorem can be derived from the results in (Pagano 1998; Slilaty and Qin 2007) and states that tangled signed graphs are precisely the graphs of binary signed-graphic matroids.

Theorem 6.7 *If Σ is a connected signed graph then $M(\Sigma)$ is binary if and only if*

(i) Σ *is tangled, or*
(ii) $M(\Sigma)$ *is graphic.*

Examples of tangled signed graphs are the graphs in Figs. 6.9 and 6.10. The connectivity of a signed graph is defined as the connectivity of its underlying graph. The next theorem shows that tangled signed graphs behave as graphs with respect to the connectivity of their corresponding matroids.

Theorem 6.8 *Let Σ be a tangled signed graph. Then Σ is 2-connected if and only if $M(\Sigma)$ is connected.*

Proof For the "only if'" part, assume that for a 2-connected tangled signed graph Σ the matroid $M(\Sigma)$ is disconnected. By Theorem 6.6, this is possible only if Σ is a B-necklace. But then Σ contains a balancing vertex and thus, Σ is not tangled which is in contradiction with our assumption.

For the "if" part, suppose that $M(\Sigma)$ is 2-connected and it does have a tangled representation Σ which is not 2-connected. Therefore Σ contains at least two blocks, and exactly one of them will be unbalanced. By Theorem 6.6 then Σ has two separates, which implies that $M(\Sigma)$ has more than one elementary separators contradicting our hypothesis about the connectivity of the matroid.   □

6.4 Decomposition

In this section we will present the results from (Papalamprou and Pitsoulis 2013) where the theory of bridges described in Chap. 5 is extended to binary signed-graphic matroids. All the definitions for binary matroids given in Sect. 5.1 regarding bridges, Y-components, separating cocircuits, and bridge-separability apply in this section also. All proofs that are not included in this section can be found in (Papalamprou and Pitsoulis 2013).

There are two main theorems in Chap. 5 which are essential for the decomposition Theorem 5.6; Theorem 5.4, which states that bridge-separability is a property of all cocircuits in graphic matroids, and Theorem 5.5, which enables us to derive a graph representation of a graphic matroid where a given cocircuit is a star of a vertex if the deletion of the cocircuit results in a set of bridges which all avoid each other. By combining these two theorems we were able to prove in Theorem 5.6 that graphic matroids admit a structural characterization, which also leads to an efficient recognition algorithm described in Sect. 5.3. It turns out that the situation is similar for signed-graphic matroids, however, due to the nature of circuits in signed graphs we have to incorporate the existence of positive and negative cycles. For a matroid M, call a cocircuit $Y \in \mathscr{C}^*(M)$ **graphic** if $M \backslash Y$ is a graphic matroid. We will show that binary signed-graphic matroids can be decomposed into graphic matroids and one non-graphic matroid with only graphic cocircuits. Furthermore, we provide an excluded minor characterization of binary matroids with only graphic cocircuits.

Recall that since we are dealing with binary signed-graphic matroids, by Theorem 6.7 we are only interested in tangled signed graphs. Therefore, the only possible circuits will be positive cycles and tight handcuffs, and the only possible bonds are balancing and unbalancing bonds. We begin with the following lemma which shows that balanced bridges of non-graphic cocircuits result in graphic minors in the signed-graphic matroid. The proof of the lemma illustrates the graphical effect of the operations of deletion and contraction when applied to bridges of non-graphic cocircuits that correspond to balanced and unbalanced subgraphs.

Lemma 6.1 *Let $M(\Sigma)$ be a binary signed-graphic matroid, Y a non-graphic cocircuit and B a bridge of Y in $M(\Sigma)$. If $\Sigma|B$ is balanced then $M(\Sigma).(Y \cup B)$ is graphic.*

Proof Since Y is a non-graphic cocircuit, $M(\Sigma)$ is not a graphic matroid, which by Theorem 6.7 it implies that Σ is a tangled signed graph. Moreover, Y will be either a star or unbalancing bond in Σ such that the core of $\Sigma \backslash Y$ is not a B-necklace. Let B^+ be any bridge of Y such that $\Sigma|B^+$ is balanced, and let B^- be the bridge corresponding to the unique unbalanced block of $\Sigma \backslash Y$. Perform switchings in the vertices of Σ such that all the edges in the balanced blocks of $\Sigma \backslash Y$ become positive.

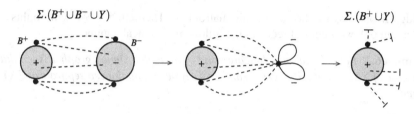

Fig. 6.11 B^+ in the balanced component of $\Sigma \backslash Y$

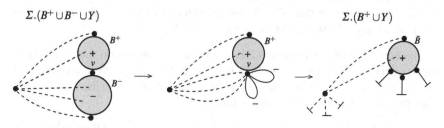

Fig. 6.12 B^+ in the unbalanced component of $\Sigma \backslash Y$

Now the bridge B^+ can be in either the balanced or the unbalanced component of $\Sigma \backslash Y$.

If B^+ is in the balanced component of $\Sigma \backslash Y$, then Y is an unbalancing bond. Contract any other balanced block to obtain $\Sigma.(B^+ \cup B^- \cup Y)$. Contracting the edges of the unbalanced block B^-, where if an edge is negative switching, one of its end-vertices will result in one or more negative loops and half-edges since this block contains negative cycles. Contraction of these negative loops and half-edges will result in the signed graph $\Sigma.(B^+ \cup Y)$ where the only negative cycles are the negative loops and half-edges of Y. The whole procedure can be seen schematically in Fig. 6.11, where the dashed lines are the edges of the bond Y and the shaded circles are the subgraphs corresponding the bridges of Y, with a sign that indicates whether the subgraph is balanced or unbalanced. Therefore, by Proposition 6.3 we have that the matroid $M(\Sigma.(B^+ \cup Y)) = M(\Sigma).(B^+ \cup Y)$ is a graphic.

If B^+ is in the unbalanced component of $\Sigma \backslash Y$, then Y can be either an unbalancing or a star bond, and the argument is similar as before. Contract again any other balanced block to obtain $\Sigma.(B^+ \cup B^- \cup Y)$ (see Fig. 6.12). Contraction now of the edges in the unbalanced block B^- will result in the edges in B^+, which are incident to the unique common vertex v of B^+ and B^-, to become negative loops and half-edges. Therefore $\Sigma.(B^+ \cup Y)$ will contain a balanced component $\bar{B}$, which is not necessarily 2-connected, and a number of negative loops and half-edges from Y and B^+. If $\Sigma.(B^+ \cup Y)$ contains a negative cycle C other than the negative loops and half-edges, then C would be a negative cycle in Σ which is disjoint from v, and therefore, vertex disjoint with any negative cycle in B^- implying that Σ is not tangled. $\qquad \square$

To establish bridge-separability for a cocircuit Y of a signed-graphic matroid $M(\Sigma)$, we need provide a graphical characterization of $\pi(M(\Sigma), B, Y)$ for any bridge B of Y, as we did for graphic matroids in Theorem 5.3. We will do this in Theorem 6.9, whose proof requires the following technical lemma.

Lemma 6.2 *Let Y be an unbalancing bond of a tangled signed graph Σ such that the core of $\Sigma \backslash Y$ is not a B-necklace and let $\Sigma | B$ be an unbalanced separate of $\Sigma \backslash Y$. Then*

(i) $M(\Sigma)$ *is graphic, or*
(ii) *there exists a series of switchings on the vertices of Σ such that all the edges of the separates other than $\Sigma | B$ become positive and the edges for all nonempty $Y(B, v)$ have the same sign for any $v \in V(\Sigma | B)$.*

The condition (ii) of Lemma 6.2 is central in the discussion that will follow, and essentially forces $Y(B, v)$ to be a bond of $\Sigma.(B \cup Y)|Y$, thus, a cocircuit in the corresponding signed-graphic matroid. This fact enables us to use Corollary 5.1 since the matroid under examination is binary.

Theorem 6.9 *Let $M(\Sigma)$ be a binary signed-graphic matroid and Y an unbalancing bond of Σ such that the core of $\Sigma \backslash Y$ is not a B-necklace. If $\Sigma | B$ is a separate of an end-graph Σ_i of $\Sigma \backslash Y$ then $\pi(M(\Sigma), B, Y)$ is the class of all nonempty $Y(B, v)$ such that $v \in V(\Sigma | B)$.*

Proof Let $\mathscr{L} = \{Y(B, v) : v \in V(\Sigma | B)$ and $Y(B, v) \neq \emptyset\}$. By Corollary 5.1, we know that

$$\pi(M(\Sigma), B, Y) = \mathscr{C}^*((M(\Sigma).(B \cup Y))|Y)$$

and since signed-graphic matroids are closed under the operations of deletion and contraction, we have

$$\pi(M(\Sigma), B, Y) = \mathscr{C}^*(M((\Sigma.(B \cup Y))|Y)).$$

Let $\mathscr{M}$ be the family of bonds of $\Sigma_d = (\Sigma.(B \cup Y))|Y$. Since there is one-to-one correspondence between the members of $\mathscr{C}^*(M((\Sigma.(B \cup Y))|Y)$ and the bonds of Σ_d, we shall equivalently show that, for any bridge B of Y in $M(\Sigma)$, $\mathscr{L} = \mathscr{M}$. This will be shown only for the case in which Y is an unbalancing bond since the proof for the case in which Y is a star bond follows easily.

The signed graph $\Sigma \backslash Y$ will consist of two components Σ_1 and Σ_2 and contain exactly one unbalanced block. Without loss of generality, we assume that this unbalanced block, say B_0, is contained in Σ_1. Since $C(B_0, v)$ is balanced for any $v \in V(\Sigma | B_0)$ and Σ_2 is balanced, there exists a series of switchings on the vertices of $\Sigma_1 \backslash B_0$ and Σ_2 such that all the edges in $\Sigma_1 \backslash B_0$ and Σ_2 become positive. We call Σ', Σ_1', Σ_2' and Σ_d' the graphs so-obtained from Σ, Σ_1, Σ_2 and Σ_d, respectively. Figure 6.13 depicts such a signed graph Σ', where negative edges will appear only on the unbalanced block B_0 and the bond Y. A bridge B of Y in $M(\Sigma') = M(\Sigma)$ can be either one of the following three cases based on the form of the corresponding separates in $\Sigma' \backslash Y$:

Fig. 6.13 Σ' for the proof of Theorem 6.9

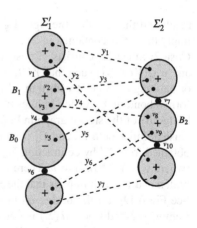

- **Case 1**, the separate $\Sigma'|B$ of $\Sigma'\backslash Y$ is a balanced block in Σ'_1,
- **Case 2**, the separate $\Sigma'|B$ of $\Sigma'\backslash Y$ is a balanced block in Σ'_2,
- **Case 3**, the separate $\Sigma'|B$ of $\Sigma'\backslash Y$ is the unbalanced block in Σ'_1.

The subgraphs B_0, B_1, and B_2 in Fig. 6.13 illustrate the above cases. In what follows we shall show that $\mathscr{L}$ is contained in $\mathscr{M}$ for any bridge B in each case.

Case 1: Let X be the set of common vertices of $\Sigma'|B$ and $\Sigma'\backslash B$ and u the vertex so obtained from contracting Σ'_2. Clearly, there exists an $v_j \in X$ such that $C(B, v_j)$ contains the unbalanced block of Σ'_1. For any $v \in V(\Sigma'|B) - \{v_j\}$ such that $Y(B, v) \neq \emptyset$, $Y(B, v)$ is a set of parallel edges incident to u and v in Σ'_d, while all the edges in $Y(B, v_j)$ are negative loops or half-edges incident with u in Σ'_d. See Fig. 6.14a for bridge B_1 of the signed graph in Fig. 6.13. Furthermore, for any $v \in V(\Sigma'|B) - \{v_j\}$ such that $Y(B, v) \neq \emptyset$, the edges of $Y(B, v)$ must be of the same sign, since otherwise Σ' would have two vertex disjoint negative cycles contradicting the fact that Σ' is tangled. Thus, any $Y(B, v) \neq \emptyset$ is a bond of Σ'_d. This

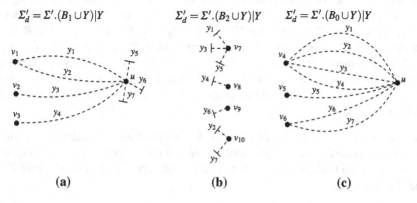

Fig. 6.14 Three different cases in the proof of Theorem 6.9. (**a**) Case 1 (**b**) Case 2 (**c**) Case 3

result and the fact that the signed graphs Σ_d and Σ'_d have equal classes of bonds imply that $\mathscr{L}$ is contained in $\mathscr{M}$.

Case 2: Since Σ'_2 consists of positive edges and Σ'_1 contains a negative cycle, for any $v \in V(\Sigma'|B)$ such that $Y(B, v) \neq \emptyset$, $Y(B, v)$ will be a set of negative loops or half-edges incident with v in $\Sigma'.(B \cup Y)$. Thus, the edges of each $Y(B, v)$ will form a bond of Σ'_d which implies that $\mathscr{L}$ is contained in $\mathscr{M}$. See Fig. 6.14b for the bridge B_2 of the signed graph in Fig. 6.13.

Case 3: Since both Σ'_2 and $\Sigma'_1 \backslash B$ consist of positive edges, the graph $\Sigma'.(B \cup Y)$ is obtained from Σ' by contracting Σ'_2 to a vertex u and by contracting each $C(B, v)$ (where $v \in V(\Sigma'|B)$) to v. Therefore, the edges of each $Y(B, v)$ become incident with u and v which implies that the edges of each $Y(B, v)$ are parallel edges in Σ'_d. See Fig. 6.14c for the bridge B_0 of the signed graph in Fig. 6.13. Furthermore, by Lemma 6.2 and since $M(\Sigma)$ is not graphic, each $Y(B, v)$ in Σ_b consists of edges of the same sign. Thus, each $Y(B, v)$ is a bond of Σ'_b which implies that $\mathscr{L}$ is contained in $\mathscr{M}$.

Finally, Σ'_d in all three cases has no other bonds, since otherwise it should have two bonds having at least one common edge. This would imply that $M(\Sigma'_d)$ would have two cocircuits which have a common element and thus, by Corollary 5.1, $M(\Sigma')$ would not be binary. By Theorem 6.7, this contradicts the fact that Σ' is tangled and thus, $\mathscr{L} = \mathscr{M}$. □

In contrast with graphic matroids, not all cocircuits of signed-graphic matroids are bridge-separable. The next theorem states that the cocircuits that correspond to the bonds described in Theorem 6.9 have the property of being bridge-separable.

Theorem 6.10 *Let Y be a cocircuit of a binary signed-graphic and non-graphic matroid $M(\Sigma)$. If Y is an unbalancing bond of Σ such that the core of $\Sigma \backslash Y$ is not a B-necklace then Y is bridge-separable.*

Proof Let Σ_1 and Σ_2 be the two distinct components of $\Sigma \backslash Y$. Arrange the bridges of Y in $M(\Sigma)$ in two classes T_1 and T_2 such that a bridge B is in T_1 or T_2 if $\Sigma|B$ is a separate of Σ_1 or Σ_2 respectively. Suppose now that two bridges B_1 and B_2 of T_1 overlap. Then $\Sigma|B_1$ and $\Sigma|B_2$ are separates of Σ_1. Thus, there exist vertices v_1 of $\Sigma|B_1$ and v_2 of $\Sigma|B_2$ such that $\Sigma|B_2$ is a subgraph of $C(B_1, v_1)$ and $\Sigma|B_1$ is a subgraph of $C(B_2, v_2)$. Furthermore, every vertex of $V(\Sigma_1)$ is a vertex of $C(B_1, v_1)$ or $C(B_2, v_2)$, therefore we have that $Y(B_1, v_1) \cup Y(B_2, v_2) = Y$. Thus, by Theorem 6.9 we can find some $K \in \pi(M(\Sigma), B_1, Y)$ and $J \in \pi(M(\Sigma), B_2, Y)$ such that $K \cup J = Y$. This contradicts our assumption that B_1 and B_2 overlap and the result follows. □

Theorem 6.11 is an extension of Theorem 5.5 to signed-graphic matroids. It shows that for any binary signed-graphic matroid M and non-graphic cocircuit Y with no two overlapping bridges there exists a signed graph representation of M where Y is the star of a vertex.

Theorem 6.11 *Let Y be a non-graphic cocircuit of a connected binary signed-graphic matroid M such that no two bridges of Y in M overlap. Then there exists a 2-connected signed graph Σ where Y is the star of a vertex and $M = M(\Sigma)$.*

The proof of Theorem 6.11 follows the lines of the proof of Theorem 5.5. It shows that the condition that any pair of bridges is avoiding induces disjoint 2-separations in Σ, and we can perform a sequence of twistings in these 2-separations to reduce one of the components of $\Sigma \backslash Y$ to a single vertex. Moreover, it is proved that these switchings satisfy condition (ii) of Proposition 6.2 so the signed-graphic matroid is not altered.

The main result in (Papalamprou and Pitsoulis 2013) is the following theorem which states that a binary signed-graphic matroid can be decomposed into minors which are all graphic, apart from one which is signed-graphic, while these conditions are also sufficient for a binary matroid to be signed-graphic.

Theorem 6.12 (Decomposition of Binary Signed-Graphic Matroids) Let M be a connected binary matroid and $Y \in \mathscr{C}^*(M)$ be a non-graphic cocircuit. Then M is signed-graphic if and only if:

(i) Y is bridge-separable, and
(ii) for any bridge B of Y, the minor $M.(B \cup Y)$ is graphic apart from one which is signed-graphic.

Proof Assume that M is signed-graphic. Since it is binary and not graphic, by Theorem 6.7, there exists a tangled signed graph Σ such that $M = M(\Sigma)$. Moreover, since $M \backslash Y$ is not graphic, Y cannot be a balancing bond or an unbalancing bond of Σ such that $\Sigma \backslash Y$ contains a B-necklace. Therefore, Y is either a star bond or an unbalancing bond such that $\Sigma \backslash Y$ does not contain a B-necklace and, by Theorem 6.10, we can conclude that Y is a bridge-separable cocircuit of M. $\Sigma \backslash Y$ will contain exactly one unbalanced block, say $\Sigma | B^-$ which is not a B-necklace, and k balanced blocks $\Sigma | B_i$ where $k \geq 0$. By Theorem 6.6, these blocks are the elementary separators of $M(\Sigma \backslash Y) = M(\Sigma) \backslash Y$, and therefore the bridges of Y in $M(\Sigma)$. By Lemma 6.1, we have that each $M(\Sigma).(B_i \cup Y)$ is a graphic matroid for each $i = 1, \ldots, k$. Since $M(\Sigma).(B^- \cup Y)$ is a minor of $M(\Sigma)$ it can be either a signed-graphic or graphic matroid. If $M(\Sigma).(B^- \cup Y)$ is graphic then Y is a bridge-separable cocircuit of a connected binary matroid such that all of its Y-components are graphic matroids, and by Theorem 5.6 we have that M is a graphic matroid, a contradiction.

The proof of sufficiency follows the argument of the proof of Theorem 5.6, and it will not be given here. The main difference here is that the matroids in first part of the proof of Theorem 5.6 are now a graphic matroid G_1 and a signed graphic matroid Σ_1 where Y is a star of a vertex in both, and an appropriate signing is performed in part two in order to create a signed graph Σ from the star composition of G_1 and Σ_1. In order to show that $M(\Sigma) = M$ the same matroid argument as in part three of the proof of Theorem 5.6 is used. □

Note that if Y is a non-separating cocircuit of M then $M \backslash Y$ has only one separate B, which implies that $M.(B \cup Y) = M$ and Theorem 6.12 holds trivially. Therefore, in order to decompose a binary matroid into proper minors based on Theorem 6.12 the existence of a cocircuit which is both separating and non-graphic is required. An excluded minor characterization of the binary signed-graphic matroids with graphic cocircuits is given by the following theorem, where the graph $K_{4,4}^-$ is $K_{4,4}\backslash e$ for any edge $e \in E(K_{4,4})$.

Theorem 6.13 *Let M be a binary matroid such that all its cocircuits are graphic. Then, M is signed-graphic if and only if M has no minor isomorphic to* $M^*(K_{3,5})$, $M^*(K_{4,4}^-)$, F_7 *or* F_7^*.

The following theorem shows that the non-existence of non-graphic separating cocircuits is a property inherited to minors created by the operation of deletion of a cocircuit.

Theorem 6.14 *If $M(\Sigma)$ is a binary signed-graphic matroid such that every non-graphic cocircuit $Y \in \mathcal{C}^*(M(\Sigma))$ is non-separating then any non-graphic cocircuit Y' of $M(\Sigma)\backslash Y$ is also non-separating.*

Proof Assume that Y' is a separating non-graphic cocircuit of $M(\Sigma)\backslash Y = M(\Sigma\backslash Y)$. Letting $\Sigma' = \Sigma\backslash Y$, and since Y' is an unbalancing bond of Σ', we have that $\Sigma'\backslash Y'$ consists of two components Σ_1' and Σ_2' which are nonempty of edges, and we may suppose that $M(\Sigma_1')$ is not graphic. Therefore, $\bar{Y} = Y' \cup S$ is a cocircuit of $M(\Sigma)$ where S is a possibly empty subset of Y. Due to the fact that Y is a star bond of Σ, all the edges in S have a common end-vertex v in Σ. Let $S_1 \subset S$ be the edges which have their end-vertices other than v in Σ_1. Then, $\hat{Y} = Y' \cup S_1$ is a minimal set of edges such that $\Sigma\backslash\hat{Y}$ consists of two components, one of which is Σ_1, where $M(\Sigma_1)$ is not graphic. This implies that $\hat{Y}$ is an unbalancing bond of Σ and therefore, $\hat{Y}$ is a separating and non-graphic cocircuit of $M(\Sigma)$; a contradiction. □

Finally, we also have the characterization of those tangled signed graphs that their corresponding singed-graphic matroids have cocircuits which are all non-graphic and non-separating.

Theorem 6.15 $M(\Sigma)$ *is a connected, binary, signed-graphic, non-graphic matroid such that every non-graphic cocircuit $Y \in \mathcal{C}^*(M(\Sigma))$ is non-separating if and only if Σ is a 2-connected signed graph such that every non-graphic cocircuit $Y \in \mathcal{C}^*(M(\Sigma))$ is a star of Σ and $\Sigma\backslash Y$ is 2-connected.*

Proof We will prove necessity first. Since $M(\Sigma)$ is binary and connected, by Theorem 6.8, Σ is tangled and 2-connected. Furthermore, any non-graphic cocircuit of $M(\Sigma)$ is non-separating and therefore, by our classification of bonds, Y is a star bond of Σ. $\Sigma\backslash Y$ has exactly two components and one unbalanced block. Moreover, $M(\Sigma)\backslash Y$ is connected since Y is non-separating and therefore, by Theorem 6.6, $\Sigma\backslash Y$ can not be a necklace or contain any other block except for the unbalanced one.

To prove sufficiency, since $M(\Sigma)$ is non-graphic, by Theorem 6.7, Σ is tangled. Furthermore, Σ is 2-connected and thus, by Theorem 6.8, $M(\Sigma)$ is connected. Any non-graphic cocircuit Y is such that $\Sigma\backslash Y$ is 2-connected; moreover, $\Sigma\backslash Y$ is not a necklace since Y is non-graphic. Thus, by Theorem 6.6, $M(\Sigma)\backslash Y$ is connected and therefore, any non-graphic cocircuit of $M(\Sigma)$ is non-separating.   □

We can combine the above results to decompose a binary signed-graphic matroid to graphic matroids and possibly one binary matroid with no $M^*(K_{3,5})$, $M^*(K_{4,4}^-)$, F_7 or F_7^* minors by successively deleting cocircuits. While there exist non-graphic separating cocircuits we apply Theorem 6.12, which dictates that the deletion of such a cocircuit will result in graphic matroids and one signed-graphic matroid $M(\Sigma)$. If all the non-graphic cocircuits of $M(\Sigma)$ are non-separating then, by Theorems 6.14 and 6.15, it is evident that all these cocircuits will correspond to stars in Σ and they can be deleted, resulting to either a graphic matroid or a signed-graphic matroid with no $M^*(K_{3,5})$, $M^*(K_{4,4}^-)$, F_7, F_7^* minors. This decomposition however does not directly lead to an algorithm for solving the recognition Problem 6.1 for binary signed-graphic matroids as it was the case for graphic matroids, since there is no efficient way of checking whether a matroid with only graphic cocircuits is signed-graphic or not. However, Papalamprou and Pitsoulis (2009) present a polynomial time algorithm to test whether a cographic matroid with graphic cocircuits is signed-graphic or not.

6.5 Notes

Signed graphs were defined by Harary (1954) and were motivated by problems in social psychology. Zaslavsky (1998) provides an extended bibliography on signed graphs. Signed-graphic matroids have been introduced by Zaslavsky (1982, 1990, 1991), and have been studied by Pagano (1998), Slilaty (2005, 2006), and Slilaty and Qin (2007) among others. Recently it has been conjectured by Mayhew et al. (2013) that they may be the building blocks of a k-sum decomposition of dyadic and near-regular matroids.

Bidirected graphs were introduced by Edmonds (1967) as a generalization of both directed and undirected graphs. Binet matrices were introduced by Appa and Kotnyek (2004, 2006), and can be viewed as generalizations of totally unimodular network matrices. Apart from their importance in optimization as stated in Theorem 6.1, they also have theoretical importance in structural results such as the decomposition theorem of totally unimodular matrices by Seymour (1980), which states that any totally unimodular matrix can be obtained by k-sums from network matrices and their transposes, and two binet matrices. Given that any network matrix is a binet matrix of a bidirected graph with only directed edges, we conclude that binet matrices are the building blocks of totally unimodular matrices. Moreover, in (Pitsoulis et al. 2009) it is shown that the k-sum between a binet and a network matrix is a binet matrix, which provides the means of constructing a bidirected graph representation of any totally unimodular matrix. A polynomial time recognition algorithm for binet matrices is given by Musitelli (2007), where in (Musitelli 2010) some of the main concepts of this recognition algorithm are presented. Papalamprou and Oitsoulis (2012) used the binet matrix recognition algorithm of Musitelli, to provide a recognition algorithm that determines whether a matroid given by an independence oracle is binary signed-graphic.

Given the representability classes of signed-graphic matroids in the beginning of Sect. 6.3 the question that naturally arises is whether it is possible to derive similar decomposition results as Theorem 6.12 for signed-graphic matroids representable over $GF(4)$, or over all fields of characteristic not two. Although this is part of an ongoing research effort, preliminary results indicate that the results of Sect. 6.4 do not seem to be readily extendible to the $GF(4)$ case. This is partly due to the fact that although for the binary case we had a relatively simple characterization of signed graphs and their matroids given by Theorem 6.7, for $GF(4)$ representable signed-graphic matroids the corresponding characterization involves more complicated signed graphs (Gerards 1990; Pagano 1998; Slilaty and Qin 2007). Moreover, for signed-graphic matroids which are not $GF(2)$ or $GF(4)$ representable, there is no such characterization.

References

1. Aigner, M.: Combinatorial Theory. Springer-Verlag, New York (1979)
2. Appa, G., Kotnyek, B.: Rational and integral k-regular matrices. Disc. Math. **275**, 1–15 (2004)
3. Appa, G., Kotnyek, B.: A bidirected generalization of network matrices. Networks **47**, 185–198 (2006)
4. Appa, G., Kotnyek, B., Papalamprou, K., Pitsoulis, L.: Optimization with binet matrices. Oper. Res. Lett. **35**, 345–352 (2007)
5. Auslander, L., Trent, H.: Incidence matrices and linear graphs. J. Math. Mech. **8**, 827–835 (1959)
6. Bixby, R.: On Reid's characterization of the ternary matroids. J. Comb. Theor. Ser. B **26**, 174–204 (1979)
7. Bixby, R., Cunningham, W.: Converting linear programs to network problems. Mathem. Oper. Res. **5**, 321–357 (1980)
8. Bixby, R., Cunningham, W.: Matroid optimization and algorithms. In: Graham, R., Grötschel, M., Lovász, L. (eds.) Handbook of Combinatorics, vol. I, chap. 11. Elsevier, Amsterdam (1995)
9. Bixby, R., Wagner, D.: An almost linear-time algorithm for graph realization. Math. Oper. Res. **13**, 99–123 (1988)
10. Bryant, V., Perfect, H.: Independence Theory in Combinatorics. Chapman Hall, London (1980)
11. Brylawski, T.: Appendix of matroid cryptomorphisms. In: White N. (ed.) Matroid Theory, Encyclopedia of Mathematics and its Applications, vol. 26. Cambridge University Press, New York (1986)
12. Crapo, H., Rota, G.C.: On The Foundations of Combinatorial Theory: Combinatorial Geometries. The MIT Press, Cambridge (1970)
13. Cunningham, W.: On matroid connectivity. J. Comb. Theor. Ser. B **30**, 94–99 (1981)
14. Cunningham, W.: Separating cocircuits in binary matroids. Linear Algebra Appl. **43**, 69–86 (1982)
15. Diestel, R.: Graph Theory. Springer, New York (2006)
16. Edmonds, J.: An introduction to matching. Technical report, The University of Michigan (Engineering Summer Conference) (1967)
17. Edmonds, J.: Submodular functions, matroids, and certain polyhedra. In: Guy, R. et al. (ed.) Combinatorial Structures and Their Applications, pp. 69–87. Gordon and Breach, New York (1970)
18. Edmonds, J.: Matroids and the greedy algorithm. Math. Program. **1**, 127–136 (1971)
19. Edmonds, J., Fulkerson, D.: Transversals and matroid partition. J. Res. Natl Bur. Stand. **69**(3), 147–153 (1965)
20. Faigle, U.: The greedy algorithm for partially ordered sets. Discrete Math. **28**, 153–159 (1979)
21. Fournier, J.: Une relation de separation entre cocircuits d'un matroide. J. Comb. Theor. Ser. B **12**, 181–190 (1974)

22. Fujishige, S.: An efficient PQ-graph algorithm for solving the graph realization problem. J. Comput. Syst. Sci. **21**, 63–86 (1980)
23. Gale, D.: Optimal assignments in an ordered set: an application of matroid theory. J. Comb. Theor. **4**, 176–180 (1968)
24. Geelen, J., Gerards, B.: Characterizing graphic matroids by a system of linear equations. Technical report, Centrum Wiskunde & Informatica (CWI). http://oai.cwi.nl/oai/asset/18866/18866B.pdf (2011)
25. Gerards, A.: Graphs and Polyhedra. Binary Spaces and Cutting Planes. CWI Tract vol. 73. Centrum voor Wiskunde en Informatica, Amsterdam (1990)
26. Gordon, G., McNulty, J.: Matroids: A Geometric Introduction. Cambridge University Press, Cambridge (2012)
27. Graham, R., Grötschel, M., Lovász, L. (eds.): Handbook of Combinatorics. Elsevier, Amsterdam (1995)
28. Hall, P.: On representatives of subsets. J. Lond. Math. Soc. **10**, 26–30 (1935)
29. Harary, F.: On the notion of balance of a signed graph. Mich. Math. J. **2**, 143–146–381 (1954)
30. Hausmann, D., Korte, B.: k-greedy algorithms for independence systems. Math. Methods Oper. Res. **22**(1), 219–228 (1978)
31. Hausmann, D., Korte, B., Jenkyns, T.: Worst case analysis of greedy type algorithms for independence systems. Math. Program. Study **12**, 120–131 (1980)
32. Helman, P., Moret, B., Shapiro, H.: An exact characterization of greedy structures. SIAM J. Discrete Math. **6**(2), 274–283 (1993)
33. Inukai, T., Weinberg, L.: Whitney connectivity of matroids. SIAM J. Algebraic Discrete Math. **2**, 108–120 (1981)
34. Jenkyns, T.: The efficiency of the greedy algorithm. In: Proceedings of the 7th Southeastern Conference on Combinatorics, Graph Theory and Computing, Utilitas Mathematica, pp. 341–350. Winipeg (1976)
35. Korte, B., Hausmann, D.: An analysis of the greedy heuristic for independence systems. Ann. Discrete Math. **2**, 65–74 (1978)
36. Korte, B., Lovász, L.: Mathematical structures underlying greedy algorithms. In: Fundamentals of Computation Theory. Lecture Notes in Computer Science, vol. 117, pp. 205–209. Springer, Berlin (1981)
37. Korte, B., Vygen, J.: Combinatorial Optimization. Theory and Algorithms, 2nd edn. Springer-Verlag, Berlin (2001)
38. Kotnyek, B.: A generalization of totally unimodular and network matrices. Ph.D. thesis, London School of Ecomonics and Political Science, London (2002)
39. Lawler, E.: Comb. Optim. Netw. Matroids. Rinehart and Winston, Holt (1976)
40. Lee, J.: A First Course in Combinatorial Optimization. Cambridge, Cambridge University Press (2004)
41. Mayhew, D., Whittle, G., van Zwam, S.: An obstacle to a decomposition theorem for near-regular matroids. SIAM J. Discrete Math. **25**, 271–279 (2013)
42. Mighton, J.: A new characterization of graphic matroids. J. Comb. Theor., Series B **98**, 1253–1258 (2008)
43. Mirsky, L.: Transversal theory and the study of abstract independence. J. Math. Anal. Appl. **25**, 209–217 (1969)
44. Mirsky, L.: Transversal Theory: an Account of Some Aspects of Combinatorial Mathematics. Mathematics in Science and Engineering, vol. 75. Academic press, New York (1971)
45. Murota, K.: Matrices and Matroids for Systems Analysis. Algorithms and Combinatorics, vol. 20. Springer-Verlag, Berlin (1999)
46. Musitelli, A.: Recognition of generalized network matrices. Ph.D. thesis, Ecole Polytechnique Federale de Lausanne (2007)
47. Musitelli, A.: Recognizing binet matrices. Math. Program. **124**, 349–381 (2010)
48. Nemhauser, G., Wolsey, L.: Integer and Combinatorial Optimization. Discrete Mathematics and Optimization. Wiley, New York (1989)
49. Nering, E.: Linear Algebra and Matrix Theory. 2nd edn. Wiley, New York (1970)

50. Nicoletti, G., White, N.: Axiom systems. In: White, N. (ed.) Matroid Theory, Encyclopedia of Mathematics and its Applications, vol. 26, chap. 2. Cambridge, Cambridge University Press (1986)

51. Ore, O.: Graphs and matching theorems. Duke Math. J. **22**, 625–639 (1955)

52. Oxley, J.: On a matroid generalization of graph connectivity. Math. Proc. Camb. Philos. Soc. **90**, 207–214 (1981)

53. Oxley, J.: Matroid Theory. Oxford, Oxford University Press (1992)

54. Oxley, J.: Matroid Theory, 2nd edn. Oxford University Press, Oxford (2011)

55. Pagano, S.: Separability and representability of bias matroids of signed graphs. Ph.D. thesis, Binghampton University (1998)

56. Papadimitriou, C., Steiglitz, K.: Combinatorial Optimization: Algorithms and Complexity. Prentice Hall, Englewood Cliffs (1982)

57. Papalamprou, K., Pitsoulis, L.: Regular matroids with graphic matroids. In Electronic Proceedings on Theoretical Computer Science, pp. 29–41. (2009)

58. Papalamprou, K., Pitsoulis, L.: Recognition algorithms for binary signed-graphic matroids. In: Giacobini, A.M., et al. (ed.) Combinatorial Optimization. Lecture Notes in Computer Science (2012)

59. Papalamprou, K., Pitsoulis, L.: Decomposition of binary signed-graphic matroids. SIAM J. Discrete Math. **27**(2), 669–692 (2013)

60. Pitsoulis, L., Papalamprou, K., Appa, G., Kotneyk, B.: On the representability of totally unimodular matrices on bidirected graphs. Discrete Math. **309**(16), 5024–5042 (2009)

61. Rado, R.: A note on independence functions. Proc. Lon. Math. Soc. **7**, 300–320 (1957)

62. Rado, R.: Note on the transinfinite case of Hall's theorem on representatives. J. Lon. Math. Soc. **42**, 321–324 (1967)

63. Rajappan, K., Stone, A.: On Okada's method for realizing cut-set matrices. J. Comb. Theor. **10**, 113–134 (1971)

64. von Randow, R.: Introduction to the Theory of Matroids. Lecture Notes in Economics and Mathematical Systems, vol. 109. Springer-Verlag, Berlin (1975)

65. Recski, A.: Matroid Theory and its Applications in Electric Network Theory and Statics. Algorithms and Combinatorics, vol. 6. Springer-Verlag, Berlin (1989)

66. Schrijver, A.: Combinatorial Optimization: Polyhedra and Efficiency. Algorithms and Combinatorics, vol. 24. Springer-Verlag, Berlin (2003)

67. Seymour, P.: Matroid representation over $GF(3)$. J. Comb. Theor. Ser. B **26**, 159–173 (1979)

68. Seymour, P.: Decomposition of regular matroids. J. Comb. Theor. Ser. B **28**, 305–359 (1980)

69. Seymour, P.: Recognizing graphic matroids. Combinatorica **1**, 75–78 (1981)

70. Seymour, P.: Matroids minors. In: Graham, R., Grötschel, M., Lovász, L. (eds.) Handbook of Combinatorics, vol. I, chap. 10. Elsevier, Amsterdam (1995)

71. Slilaty, D.: On cographic matroids and signed-graphic matroids. Discrete Math. **301**, 207–217 (2005)

72. Slilaty, D.: Bias matroids with unique graphical representations. Discrete Math. **306**, 1253–1256 (2006)

73. Slilaty, D., Qin, H.: Decompositions of signed-graphic matroids. Discrete Math. **307**, 2187–2199 (2007)

74. Tamari, R.: Combinatorial algorithms in certain classes of binary matroids. Ph.D. thesis, Cornell University, Ithaca, New York (1977)

75. Truemper, K.: Matroid Decomposition. Academic Press, Boston (1992)

76. Tutte, W.: A class of abelian groups. Can. J. Math. **8**, 13–28 (1956)

77. Tutte, W.: A homotopy theorem for matroids I. Trans. Am. Math. Soc. **88**, 144–160 (1958a)

78. Tutte, W.: A homotopy theorem for matroids II. Trans. Am. Math. Soc. **90**, 161–174 (1958b)

79. Tutte, W.: Matroids and graphs. Trans. Am. Math. Soc. **90**, 527–552 (1959)

80. Tutte, W.: An algorithm for determining whether a given binary matroid is graphic. Proc. Am. Math. Soc. **11**, 905–917 (1960)

81. Tutte, W.: Lectures on matroids. J. Res. Natl Bur. Stand. B **69**, 1–47 (1965)

82. Tutte, W.: Connectivity in matroids. Can. J. Math. **18**, 1301–1324 (1966)

83. Tutte, W.: Introduction to the Theory of Matroids. Elsevier Science, Amsterdam (1971)
84. Welsh, D.: Generalized versions of Hall's theorem. J. Comb. Theor. Ser. B **10**, 95–101 (1971)
85. Welsh, D.: Matroid Theory. Academic Press, London (1976)
86. Welsh, D.: Matroids: fundamental concepts. In: Graham, R., Grötschel, M., Lovász, L. (eds.) Handbook of Combinatorics, vol. I, chap. 9. Elsevier, Amsterdam (1995)
87. White, N. (ed.): Theory of Matroids, Encyclopedia of Mathematics and its Applications, vol. 26. Cambridge University Press, New York (1986)
88. White, N. (ed.): Combinatorial Geometries. Encyclopedia of Mathematics and Its Applications, vol. 29. Cambridge University Press, Cambridge (1987)
89. White, N. (ed.): Matroid Applications. Encyclopedia of Mathematics and its Applications, vol. 40. Cambridge University Press, New York (1992)
90. Whitney, H.: 2-isomorphic graphs. Am. J. Math. **55**, 245–254 (1933)
91. Whitney, H.: On the abstract properties of linear dependence. Am. J. Math. **57**, 509–533 (1935)
92. Wilson, R.: An introduction to matroid theory. Am. Math. Monthly **80**(5), 500–525 (1973)
93. Wilson, R.: Introduction to Graph Theory. Addison Wesley Longman, Harlow (1996)
94. Zaslavsky, T.: Signed graphs. Discrete Appl. Math. **4**, 47–74 (1982)
95. Zaslavsky, T.: Biased graphs whose matroids are special binary matroids. Graphs Combin. **6**, 77–93 (1990)
96. Zaslavsky, T.: Biased graphs. II. The three matroids. J. Comb. Theor. Ser. B **51**, 46–72 (1991)
97. Zaslavsky, T.: A mathematical bibliography of signed and gain graphs and allied areas. Electron. J. Combin. http://cs.anu.edu.au/publications/eljc/Surveys/ds8.pdf. Dynamic Survey 8 (1998)
98. Zaslavsky, T.: A simple algorithm that proves half-integrality of bidirected network programming. Network **46**, 36–38 (2006)

Index

Symbols
Y-component, 76
k-connectivity
    in graphs, 8
    matroids, 71
k-partite, 7
k-separation
    in graphs, 8
    matroids, 71

A
Adjacent
    edges, 6
    vertices, 6
Algorithm
    GRAPHIC, 88
    GREEDY, 39
Arc, 7
Avoiding bridges, 78

B
B-necklace, 108
Balanced, 102
Balancing bond, 107
Balancing vertex, 102
Base, 27
Basis, 14
Bidirected graph, 102
Binary field, 3
Binary matroid, 27
Binet matrix, 104
Bipartite, 7
Block, 108
    inner, 108

outer, 108
Bond
    of a graph, 8
    of a signed graph, 107
Bridge, 76
Bridge-separable cocircuit, 78

C
Circuit
    of a signed graph, 106
    definition, 29
    fundamental, 31
    transversals, 21
Closed set, 36
Closure, 34
Cobase, 53
Cocircuit, 53
Cographic, 55
Column space, 4
Compact representation matrix, 49
Complete, 6
Connected
    graph, 8
Contraction
    elements in a matroid, 59
    columns in a matrix, 65
    edges in a graph, 6
    edges in a signed graph, 101
Corank, 53
Core, 108
Cryptomorphisms, 45
Cut, 6
Cycle, 7
Cycle matroid, 27

L. S. Pitsoulis, *Topics in Matroid Theory*,
SpringerBriefs in Optimization, DOI: 10.1007/978-1-4614-8957-3,
© Leonidas S. Pitsoulis 2014